# SpringerBriefs in Computer Science

*Series Editors*
Stan Zdonik
Peng Ning
Shashi Shekhar
Jonathan Katz
Xindong Wu
Lakhmi C. Jain
David Padua
Xuemin Shen
Borko Furht
VS Subrahmanian
Martial Hebert
Katsushi Ikeuchi
Bruno Siciliano

T0214149

For further volumes:
http://www.springer.com/series/10028

SpringerBriefs in Computer Science

Han Zhao • Xiaolin Li

# Resource Management in Utility and Cloud Computing

Han Zhao
University of Florida
Gainesville, FL, USA

Xiaolin Li
University of Florida
Gainesville, FL, USA

ISSN 2191-5768  ISSN 2191-5776 (electronic)
ISBN 978-1-4614-8969-6  ISBN 978-1-4614-8970-2 (eBook)
DOI 10.1007/978-1-4614-8970-2
Springer New York Heidelberg Dordrecht London

Library of Congress Control Number: 2013950917

Printed on acid-free paper

Springer is part of Springer Science+Business Media (www.springer.com)

*To our families for all their constant love, support, and encouragement.*

# Preface

The research of distributed systems encompasses many areas of computer science and engineering and is among the fastest growing subjects in the last decade. The computing trend that processor technology, driven by the Moor's Law, hits performance and power walls, and cloud computing permeates into every aspect of our life, dramatically changes the landscape of computing, making parallel and distributed computing a key norm of the computer science and engineering discipline. There is a paradigm shift that local computers no longer have to do all the heavy lifting. Instead, due to the technology advance of networking and virtualization, applications become "services" that can be purchased on demand. The emerging cloud computing model leverages remote hardware and software resources, and fuels rapid application development innovation with highly scalable and efficient computing infrastructure. Since cloud features resource provisioning elasticity, computing and storage can be packaged as metered services, known as utility computing. There is no doubt that cloud and utility computing will be one of the key driving forces to transform the entire computing industry.

Designing efficient resource management strategies is among the key issues in a cloud and utility computing environment. The introduction of socioeconomic approaches into distributed computing research opens tremendous research opportunities. This book presents cost-effective resource management strategies in cloud and utility computing, based on Dr. Han Zhao's doctoral research (*Zhao, H: Exploring Cost-Effective Resource Management Strategies in the Age of Utility Computing. Ph.D. dissertation, Department of Computer and Information Science and Engineering, University of Florida (2013)*). We have further extended the content to cover additional aspects in the field that we feel are relevant and interesting. We hope this book will help facilitate your understanding of this interesting subject.

Gainesville, FL, USA

Han Zhao
Xiaolin Li

# Acknowledgments

We would like to express our sincere thanks and appreciation to the people at Springer, USA, and in particular to Susan Lagerstrom-Fife and Courtney Clark, for their generous help throughout the publication preparation process.

We also appreciate the support from the National Science Foundation via grants OCI-1245880, OCI-1229576, CCF-0953371, CCF-1128805, OCI-0904938, CNS-091639, and CNS-0709329.

# Contents

# Chapter 1
# Introduction

**Abstract** We are entering an era of "Everything-as-a-Service" where resources are shared at an unprecedented scale. The so called utility computing model, built upon cloud computing infrastructures, becomes ubiquitous in the enterprise IT landscape. In this chapter, we first introduce recent advances in the study of the economics of cloud computing, known as Cloudnomics. Next, we describe the motivation behind cost-effective resource management design in Cloudnomics. We also summarize relevant research to the study of resource management in utility and cloud computing. At the end of this chapter, we describe the fundamental research challenges, present our design evolutions with regard to these challenges, and sketch our proposed solutions.

## 1.1 Resource Management in Utility and Cloud Computing

Distributed computing paradigms have undergone profound changes in the past decade. The emerging cloud computing [6, 14, 21, 23, 49, 54] and utility computing [42, 43] model promise to deliver agile, metered computing services to both business and scientific communities. To cope with the change, it is of paramount importance to develop efficient and flexible resource management strategies. However, the problem of managing resource allocations in a cloud and utility computing environment is challenging because both resources and administrative parties who operate these resources feature diverse heterogeneity. As cloud proliferates, scalable resource sharing platform instantiated on multiple resource providers becomes cheaper and more accessible. Hence, strategy design for resource management should equally address the heterogeneous interests of various involved parties who pursue maximum economic benefits. As a result, an inter-disciplinary research approach that combines economic models in social computing scenarios with algorithmic design in computer science becomes a viable option for researchers to build cost-effective resource scheduling strategies.

H. Zhao and X. Li, *Resource Management in Utility and Cloud Computing*,
SpringerBriefs in Computer Science, DOI 10.1007/978-1-4614-8970-2_1,
© The Author(s) 2013

### 1.1.1 Cloudnomics: Discover Economics in Cloud Computing

The fast development of parallel and distributed computing paradigms, driven by the increasing demand for computing power and network bandwidth, spurs the development of a variety of massively distributed computing platforms, such as Peer-to-Peer (P2P), Cluster, and Grid Computing emerging in academic and industrial communities. Most recently, we have witnessed the commoditization of computing, storage, network, and software: resources alone have become the subject of commercial transactions. The emerging Cloud Computing paradigm is gaining momentum as the industry realizes the economic benefits. As estimated by Gartner® [26], more than 50% of Global 1,000 companies will have stored customer-sensitive data in the public cloud by year-end 2016. For big players, cloud computing presents new opportunities to take advantage of their technology edge on computing infrastructure and IT services, allowing them to build ecosystems in the application world. For small companies, adopting cloud computing librates them from purchasing and maintaining resources, therefore greatly improves productivity and reduces time to market.

As cloud computing represents a big leap forward for distributed computing, it also brings new varieties to the design space. Performance still matters, but more importantly, it is performance per dollar that matters the most in cloud and utility computing. A new concept called Cloudnomics [18, 50] emerges from the cloud marketplace, concerning about the economics of the cloud. Essentially, lowering the cost of deploying new services and applications to the cloud is of every bit as much value as meeting service demand for organizations transiting to cloud. When it comes to research, Cloudnomics opens up many possibilities to apply economic and game theories to the scheduling and management of cloud resources. In this book, we present our research findings for cost-effective resource management strategy design in cloud and utility computing.

### 1.1.2 Motivation

Resource management, which concerns the efficient and effective acquisition and deployment of computational, storage and networking resources, is among the most important research topics in utility and cloud computing. Conventional methods for resource management, mostly centralized, are difficult to adapt to the growing complexity of modern heterogeneous distributed environments. This growing complexity is mainly caused by two phenomena. First, the distributed system becomes more loosely coupled, as its scale increases dramatically in the past few years. Today's distributed computing platform grows from clusters in a single laboratory to multiple geographically distributed computational sites, each of them composed of hundreds of computational nodes and featuring high autonomy. Second, computational devices become more heterogeneous than ever before. Many types of devices nowadays are capable of offering computing power that is only available on supercomputers decade ago, including tablet computers, smartphones, gaming consoles, etc. As a result, the increasing scalability and heterogeneity bring many challenges to effective resource management strategy design.

To address these challenges, researchers have proposed many resource management solutions to tackle the growing complexity, both theoretically and practically. Among them, one type of approach is to use market oriented mechanisms to regulate the scheduling decision making process. The market oriented mechanisms regard distributed computing units as economically rational individuals in the human society, and characterize the cooperation and competition process using economic theory. Such a socioeconomic approach most naturally models the system since it precisely capture the essential features of the modern heterogeneous distributed environments, and scales well to larger and more complex loosely coupled distributed platforms.

The utility computing model offers rental-based access to a massive pool of computational power and provides metering and accounting functions for resource usage. This service provisioning model has quite a long history (e.g., [25]), but only becomes popular as cloud computing prevails. Cloud computing eliminates the heavy economic burden of resource setup and operational cost for industry, and liberates the software development process by offering on-demand service provisioning. The utility computing model is mostly built on a cloud-based infrastructure, and defines accounting policies for resource acquisition. It helps users gain access to computational resources at a tremendous scale. When we step into the age of cloud and utility computing, we expect to see new computing paradigms implementing "Everything-as-a-Service" and to be charged at reasonable rate. Most importantly, cost-effective resource management strategies are highly desirable so that resources are utilized at low cost, and as efficiently as possible. Again, the increasing scalability and heterogeneity present tremendous challenges, as one needs to cope with varying provider-side resource availability and pricing, as well as fluctuating user-side application configuration and demand. Especially, we recognize three fundamental issues that govern the exploration in cost-effective resource management strategies in this book.

1. The flourish of virtualization technology enables more flexible resource aggregation and presents an exponential search space for optimization.
2. The heterogeneous nature of user interests has direct impact on resource management decisions.
3. Financial cost plays an important role in determining the achievable application performance.

To address these issues, we develop several resource management strategies that achieve cost-effectiveness and flexibility with regard to various scheduling contexts in cloud and utility computing. The research presented in this book highlights these challenges and provides a set of possible solutions for cost-effective resource management. Our study seeks to investigate economic models and their implication to the utility-oriented scheduling problems.

## 1.2 Summary of Related Research

In this section, we briefly summarize prior literature related to the research findings presented in this book.

### 1.2.1 Applying Socioeconomic Approach to Utility and Cloud Computing

In recent years, we have witnessed a burst of research efforts that study the application of economic and game theory for resource management in heterogeneous distributed systems [2,5,10,12,14,27,30]. An early work proposing computational economy for resource scheduling was presented in [51]. The authors presented two different market strategies for grid computing, namely commodities market and auction. The advantage of using auctions in a computational economy is that it's beneficial to discover commodity prices through strategic mechanism design. This trend is largely contributed to the following observations: design similarity of economic market mechanisms and distributed system scheduling principles; and role similarity of realistic rational individuals and egocentric heterogeneous computers. Therefore, market-oriented methods derived from game theory are extremely helpful in modeling behaviors of benefit-driven agents. Methods using game theory converge to system equilibrium state on the basis of revenue maximization. The key challenge is to identify a suitable objective function that defines target performance optimization in term of utility. Example applications of economic methods have been proposed for various scheduling topics including but not limited to dynamic resource sharing [44], workload balancing [22,31] and promoting incentives in grid and P2P systems [40]. The recent development of cloud computing technologies also urges researchers to investigate the application of auctions to manage and schedule cloud resources [13,24,53].

Depending on assumptions on individual computing resource contributor, the market oriented methods can be categorized as cooperative or noncooperative. The cooperative methods [41] take advantage of cooperative behaviors of individual sites for optimal performance scheduling, while the noncooperative methods [32, 37] explore the inherent self-interested nature of computer peers and design negotiation strategies towards utility maximization. As the distributed computing platforms become more loosely coupled, we envision that hybrid architecture is more suitable to model modern distributed systems, where cooperation and non-cooperation coexist at various scheduling and management levels. This is well conformed to distributed computing platforms at global scale, e.g., P2P desktop grid systems such as Cohesion [45] and OurGrid [17]. Further, we argue that current literature of market oriented approaches suffers from the following problems: (1) inaccurate modeling of egocentric agent behaviors; (2) lack of efficient resource management mechanism design for multi-criteria optimization; and (3) insufficient research investigation on ideas from other disciplines, mainly from economic and financial fields. In this book, we strive to address these issues in our research for cost-effective resource management strategy design.

## 1.2.2  Cost-Benefit Analysis for Utility and Cloud Computing

Cloud computing has drawn significant attentions from the industry as well as the scientific community in the past a few years. We introduce some representative examples in this section. A number of prior works [9,20,48] have attempted to study the cost-benefit of running computational and data intensive applications in cloud. For instance, Iosup et al. [34] analyzed the performance of cloud computing services for scientific computing workloads and evaluated the cost models in popular commercial cloud computing platforms. Kondo et al. [35] compared the performance and monetary cost-benefits of clouds for desktop grid applications, ranging in computational size and storage. Assunção et al. [7] conducted a cost-benefit analysis of using cloud computing to extend the capacity of clusters. In addition to computational cost, moving and storing large data set in cloud also incurs huge cost comparable [20]. Therefore, application service providers are highly motivated to carefully plan for resource usage in accordance with the estimated workload. For resource rental planning in cloud markets, various optimization models based on (non-)linear programming were proposed. For example, Goudarzi and Pedram [28] formulated the multi-dimensional SLA-based resource allocation problem as a mixed integer non-linear programming problem, and provided a heuristic solution based on force-directed scheduling. Qian and Medhi [39] presented an optimization model for minimizing server operational cost in data centers. We will propose a novel resource rental planning design in Chap. 2.

## 1.2.3  Merging Peer-to-Peer Computing with Cloud Computing

Cloud computing is promising in meeting the ever-increasing computational requirements. However, it is confronted with significant challenges on increasing scalability and complexity. One promising solution is to utilize the untapped resources interconnected in a Peer-to-Peer (P2P) manner. The idea of leveraging P2P technologies is not new. The so-called volunteer computing model [3], uses cycle scavenging and low-end desktop machines for computationally intensive applications. Compared to a centralized model, the P2P-based desktop grid computing systems achieves low cost and more flexibility, making it a good candidate for processing embarrassingly parallel computational workloads. Example projects include Condor [38], BOINC [4], Entropia [15], and GridSAT [16]. Interested readers can refer to our survey work in this field [57]. Cloud computing can also inherit the merits of P2P technologies and achieve higher flexibility for resource allocation. A few P2P cloud systems have been proposed to avoid centralized bottleneck in cloud storage [52], and to gain more flexibility in task and resource management [8, 29]. A P2P communication infrastructure is also suitable for resource-constrained mobile devices better utilize cloud services [33, 36]. The lack of centralized coordination presents many challenges in building a P2P cloud system. We will discuss relevant issues and propose a resource trading framework in Chap. 3.

### 1.2.4   Cloudnomics in Practice

Due to the service-oriented paradigm shift, utility and cloud computing revolutionize the way IT services are delivered. There are several projects (e.g., [11, 47]) that attempt to build a cloud economy. The economy allows users to bid for resource quotas, or bundles of resources for long-term use. SpotCloud™ [46] is a market place for cloud capacity trading. Some projects aim to realizing a virtual computational cloud across multiple administrative domains, e.g., CometCloud [19] and 4CaaSt [1]. In Chap. 4, we present a novel resource rental and lease platform called CloudBay. It is designed to bridge the scattered scientific communities in support of efficient running of High Performance Computing (HPC) applications.

## 1.3   Contributions

In this section, we briefly summarize the research contributions presented in this book. We start from a list of challenges for designing cost-effective resource management strategies in utility and cloud computing. We then describe the evolution of our design and methodologies to overcome these challenges. We finally present a solution summary and the organization of this book to facilitate the exploration of the readers.

### 1.3.1   Challenges

Similar to the motto for the Olympics known as "Faster, Stronger, Higher", the trend for building loosely coupled distributed systems is "Faster, Larger, Cheaper" that strive to keep up with the increasing volume of data. Many distributed computing service providers offer global span infrastructures featuring long-haul networks and multi-site coordination. As a result, it is harder to maintain homogeneity within the system. New challenges for managing resource allocation rise when heterogeneity exists and affects scheduling. Here we present three challenges to resource management in utility and cloud computing. For each challenge listed below, we describe an example scenario that will be addressed in later chapters.

1. *Challenge*: The optimization complexity is greatly increased when cost factor comes into account.

   *Scenario*: Data transfer service is priced at $x$ dollars per GB and storage service is priced at $y$ dollars per GB·month.
   *Question*: Should we cache the input data or transfer input on demand in order to provide best cost-effective computing service?

2. *Challenge*: Resource demands critical to allocation decisions are fluctuating all the time.

   *Scenario*: Alice reserves a number of VM instances for some project using reservation pricing with 1-year contract, but the project is canceled after 3 months.
   *Question*: How to cut monetary loss caused by the unexpected event? Or better yet, make profit from it?

3. *Challenge*: The flourish of cloud services present exponential combination of choices to end users.

   *Scenario*: Bob can deploy his application service on $A$ servers of type $B$ from resource provider $C$, or on $X$ servers of type $Y$ from resource provider $Z$.
   *Question*: Which option gives Bob better cost-effective results?

## 1.3.2  Design Evolution

Consider a resource customer wishing to obtain a set of computational resources from a publicly accessible resource pool, as illustrated in Fig. 1.1. One of the major challenges is how to meet the computational service demand while at the same time, reduce resource acquisition cost. We model the resource customer to be utility-driven, i.e., they are interested in profitable trading activities and independently make management decisions. This section summarizes our design evolution of cost-effective resource management strategies for the utility-driven resource customer.

The first resource management strategy design presented in this book focuses on interactions between the resource customer and the resource provider, as illustrated in Fig. 1.2. In particular, we focus on optimal strategy design for utility-oriented cloud resource rental planning. Cloud computing revolutionizes the use and deployment of IT services and spurs the emergence of Application Service Providers (ASPs) who provide managed application hosting services using cloud resources. With the knowledge of resource pricing options, a major issue faced by ASPs is how to intelligently plan resource usage for a certain time horizon in order to minimize rental cost while meeting the projected demand schedule. We found that little work has been devoted to leverage application elasticity (through job spawning and migration) to lower resource rental cost [58]. What's more interesting is that resource pricing can be dynamic (e.g., as we see in Amazon®'s spot instance

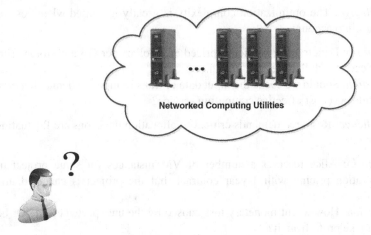

**Fig. 1.1** A user facing resource acquisition and management challenges

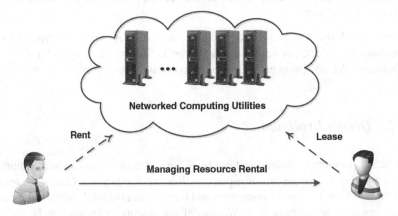

**Fig. 1.2** Design revolution 1: managing resource rental

market), making it more difficult to choose the best resource rental option under uncertainties of resource availability. We highlight these challenges and present our design in Chap. 2.

The second design takes one step further, presenting an efficient and fair resource trading framework among resource customers, as depicted in Fig. 1.3. We propose a set of utility-based multitenancy negotiation protocols to facilitate resource trading activities. In a multi-tenant environment, it is critical to satisfy different user interests in a fair, manageable, and productive way. Our research on this topic [56] uses a multi-agent approach that: (1) models tenants as utility-driven, intellectual individuals; (2) quantifies allocation benefit and loss using well-designed valuation functions; and (3) establishes a resource trading framework that evolves towards better allocation state when agents only make self-benefit trade decisions.

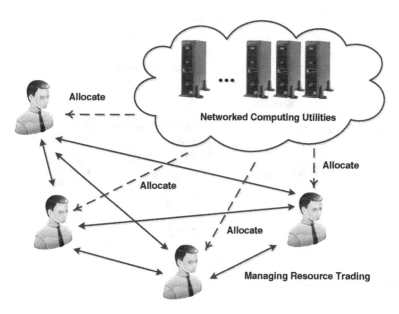

**Fig. 1.3** Design revolution 2: managing resource trading

Finally, we design CloudBay [59], a cloud-based middleware platform for cross-domain resource sharing. This design aims at providing a resource trading platform that let people rent and lease computational resources just like they buy and sell their common commodities on eBay. It features a flexible architecture in which privately owned resources form the networked computing utilities, as illustrated in Fig. 1.4. Therefore, it is most challenging to our research on cost-effective resource management strategy design. We present the prototype design of CloudBay as our initial efforts of implementing High Performance Computing (HPC)-as-a-Service. CloudBay deploys HPC and cloud services (e.g., MPI, Hadoop) on dedicated hosts contributed by scientific communities. Similar to the PlanetLab project, it provides purpose-built software from ground up, including an operating system. Prepackaged software services are encapsulated in virtual containers called Cloud Appliances. Two important services are considered as essential building blocks to enable CloudBay's functionalities: (1) autonomic networking service, which offers labor-free resource bundling based on a virtualized P2P networking library; and (2) preemptive job scheduling service, which uses Condor to manage job submission, checkpointing and preemption.

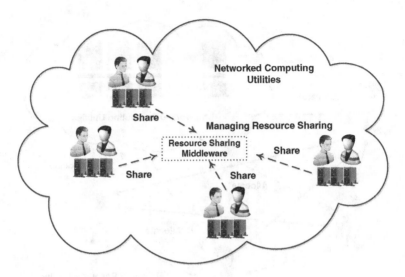

**Fig. 1.4** Design revolution 3: managing resource sharing

## *1.3.3 Summary of Solutions*

### 1.3.3.1 An Optimization Strategy Design for Cloud Resource Rental Planning

We conduct a thorough investigation of the cloud resource rental planning problem. Modeling as a mixed integer linear program, we solve the optimal resource rental problem under deterministic pricing constraints. Further, we run time series analysis on the complete history of Amazon®'s spot price variations, and propose an alternative solution that applies a stochastic optimization approach (multistage recourse) to the resource rental planning problem in order to cope with the stochastic pricing challenge. Through thorough investigation which uses Amazon®'s EC2 market as a case of study, we conclude that we cannot count on prediction results since the data correlation is weak. On the contrary, the stochastic optimization approach is more effective in hedging against pricing uncertainties. Combining the deterministic and stochastic optimization approaches, this work presents empirical values for ASPs to deploy cost-effective application services in cloud.

### 1.3.3.2 An Efficient and Fair Resource Trading Framework for Community Cloud

Inspired by previous studies in artificial intelligence, we present a unified model that uses a directed hypergraph to simultaneously capture resource allocation efficiency and unbalance amongst agents. When budget constraint presents, we propose a set of heuristic algorithms that work on distributed environment and guide agents to spontaneously trade resources. Through theoretical analysis, we show that these trading activities will improve the resource allocation within the community cloud. This resource trading framework is designed to enhance resource management in a highly collaborative resource sharing environment. For example, enabling resource trading in popular cloud resource market such as Amazon® EC2 will grant customers with "subleasing" rights, making it more attractive to cloud customers wishing a more cost-effective resource provisioning solution.

### 1.3.3.3 A Cloud Middleware for Scalable Resource Sharing

We propose a novel system design, CloudBay, which is a flexible resource management middleware to accommodate both quality-sensitive and cost-sensitive service requests. CloudBay's service scheduling scheme is inspired by eBay's transaction model, where customers can choose to "buy-it-now" or bid for an item. Therefore, CloudBay is viewed as the eBay of scientific computing resources that allows researchers to rent or release resources in a global research cloud just like they buy and sell daily commodities on eBay. Depending on user's willingness to pay (WTP), CloudBay prioritizes user service requests and schedule them accordingly. We propose a novel auction form [55] that uses proxy iterative bidding to achieve efficient auction outcome. The proposed utility model maintains incentive compatibility as in VCG auction, and is computationally tractable in winner determination. We validate the prototype of CloudBay across a variety of open and private cloud platforms, including university clusters, FutureGrid, and Amazon® EC2. Results show that CloudBay makes good use of idle resources and provides easy-to-access resources to researchers in a fair manner.

## 1.4 Organization

The contents of this book are organized as follows. In Chap. 2, we propose a fine-grained resource rental planning design for both flat-rate and spot markets. In Chap. 3, we formulate a resource trading framework for a community-based cloud computing environment. Chapter 4 describes our proof-of-concept design and implementation of a resource sharing market place. Finally, in Chap. 5, we summarize our research findings and discuss future research directions.

## References

1. 4CaaSt. Available: http://4caast.morfeo-project.org/ (2012)
2. Amar, L., Mu'alem, A., Stößer, J.: On the importance of migration for fairness in online grid markets. In: Proceedings of the 7th international joint conference on Autonomous agents and multiagent systems (AAMAS 08), pp. 1299–1302 (2008)

3. Anderson, D., Fedak, G.: The computational and storage potential of volunteer computing. In: Sixth IEEE International Symposium on Cluster Computing and the Grid (CCGRID'06)., vol. 1, pp. 73–80 (2006)

4. Anderson, D.P.: BOINC: a system for public-resource computing and storage. In: Proceedings of the Fifth IEEE/ACM International Workshop on Grid Computing, pp. 4–10 (2004)

5. Anshelevich, E., Dasgupta, A., Kleinberg, J., Tardos, E., Wexler, T., Roughgarden, T.: The price of stability for network design with fair cost allocation. In: Proceedings of the 45th Annual IEEE Symposium on Foundations of Computer Science (FOCS 04), pp. 295–304 (2004)

6. Armbrust, M., Fox, A., Griffith, R., Joseph, A.D., Katz, R., Konwinski, A., Lee, G., Patterson, D., Rabkin, A., Stoica, I., Zaharia, M.: A view of cloud computing. Commun. ACM **53**(4), 50–58 (2010)

7. Assunção, M.D., Costanzo, A., Buyya, R.: A cost-benefit analysis of using cloud computing to extend the capacity of clusters. Cluster Computing **13**(3), 335–347 (2010)

8. Babaoglu, O., Marzolla, M., Tamburini, M.: Design and implementation of a p2p cloud system. In: Proceedings of the 27th Annual ACM Symposium on Applied Computing (SAC'12)., pp. 412–417 (2012)

9. Berriman, G.B., Deelman, E., Juve, G., Regelson, M., Plavchan, P.: The application of cloud computing to astronomy: A study of cost and performance. CoRR (2010)

10. Brynjolfsson, E., Hofmann, P., Jordan, J.: Cloud computing and electricity: beyond the utility model. Commun. ACM **53**(5), 32–34 (2010)

11. Buyya, R.: Economic-based Distributed Resource Management and Scheduling for Grid Computing. CoRR **cs.DC/0204048** (2002)

12. Buyya, R., Abramson, D., Giddy, J.: Nimrod/g: an architecture for a resource management and scheduling system in a global computational grid. In: Proceedings of the fourth International Conference/Exhibition on High Performance Computing in the Asia-Pacific Region, pp. 283–289 (2000)

13. Buyya, R., Pandey, S., Vecchiola, C.: Cloudbus toolkit for market-oriented cloud computing. In: Proceedings of the 1st International Conference on Cloud Computing (CloudCom'09), pp. 24–44 (2009)

14. Buyya, R., Yeo, C.S., Venugopal, S.: Market-oriented cloud computing: Vision, hype, and reality for delivering it services as computing utilities. In: 10th IEEE International Conference on High Performance Computing and Communications (HPCC '08), pp. 5–13 (2008)

15. Chien, A.A., Calder, B., Elbert, S., Bhatia, K.: Entropia: architecture and performance of an enterprise desktop grid system. J. Parallel Distrib. Comput **63**(5), 597–610 (2003)

16. Chrabakh, W., Wolski, R.: GridSAT: a system for solving satisfiability problems using a computational grid. Parallel Computing **32**(9), 660–687 (2006)

17. Cirne, W., Brasileiro, F., Andrade, N., Costa, L., Andrade, A., Novaes, R., Mowbray, M.: Labs of the world, unite!!! Journal of Grid Computing **4**(3), 225–246 (2006)

18. Cloudonomics: The Economics of Cloud Computing. Rackspace Hosting

19. CometCloud. Available: http://nsfcac.rutgers.edu/CometCloud/ (2010)

20. Deelman, E., Singh, G., Livny, M., Berriman, B., Good, J.: The cost of doing science on the cloud: the montage example. In: Proceedings of the 2008 ACM/IEEE conference on Supercomputing (SC '08) (2008)

21. Dikaiakos, M., Katsaros, D., Mehra, P., Pallis, G., Vakali, A.: Cloud computing: Distributed internet computing for it and scientific research. Internet Computing, IEEE **13**(5), 10–13 (2009)

22. Ferguson, D., Yemini, Y., Nikolaou, C.: Microeconomic algorithms for load balancing in distributed computer systems. In: 8th International Conference on Distributed Computing Systems, pp. 491–499 (1988)

23. Foster, I., Zhao, Y., Raicu, I., Lu, S.: Cloud computing and grid computing 360-degree compared. In: Grid Computing Environments Workshop, 2008 (GCE '08), pp. 1–10 (2008)

24. Fujiwara, I., Aida, K., Ono, I.: Applying double-sided combinational auctions to resource allocation in cloud computing. In: 10th IEEE/IPSJ International Symposium on Applications and the Internet (SAINT'10), pp. 7–14 (2010)

25. Garfinkel, S., Abelson, H.: Architects of the Information Society: 35 Years of the Laboratory for Computer Science at MIT. Architects of the Information Society: 35 Years of the Laboratory for Computer Science at MIT. Mit Press (1999)
26. Gartner.http://www.businesswire.com/news/home/20111201005541/en/Gartner-Reveals-Top-Predictions-Organizations-Users-2012
27. Ghosh, P., Roy, N., Das, S., Basu, K.: A game theory based pricing strategy for job allocation in mobile grids. In: 18th International Parallel and Distributed Processing Symposium (IPDPS'04). (2004)
28. Goudarzi, H., Pedram, M.: Multi-dimensional sla-based resource allocation for multi-tier cloud computing systems. In: IEEE Cloud 2011 (2011)
29. Graffi, K., Stingl, D., Gross, C., Nguyen, H., Kovacevic, A., Steinmetz, R.: Towards a p2p cloud: Reliable resource reservations in unreliable p2p systems. In: IEEE 16th International Conference on Parallel and Distributed Systems (ICPADS'10)., pp. 27–34 (2010)
30. Grosu, D., Chronopoulos, A.: Algorithmic mechanism design for load balancing in distributed systems. IEEE Transactions on Systems, Man, and Cybernetics, Part B: Cybernetics $34$(1), 77–84 (2004)
31. Grosu, D., Chronopoulos, A.T.: Noncooperative load balancing in distributed systems. J. Parallel Distrib. Comput. $65$(9), 1022–1034 (2005)
32. Han, Z., Liu, K.J.R.: Noncooperative power-control game and throughput game over wireless networks. IEEE Transactions on Communications $53$(10), 1625–1629 (2005)
33. Huerta-Canepa, G., Lee, D.: A virtual cloud computing provider for mobile devices. In: Proceedings of the 1st ACM Workshop on Mobile Cloud Computing & Services: Social Networks and Beyond (MCS'10), pp. 6:1–6:5 (2010)
34. Iosup, A., Ostermann, S., Yigitbasi, N., Prodan, R., Fahringer, T., Epema, D.: Performance analysis of cloud computing services for many-tasks scientific computing. IEEE Trans. Parallel Distrib. Syst. $22$(6), 931–945 (2011)
35. Kondo, D., Javadi, B., Malecot, P., Cappello, F., Anderson, D.P.: Cost-benefit analysis of cloud computing versus desktop grids. In: Proceedings of the 2009 IEEE International Symposium on Parallel&Distributed Processing (IPDPS'09), pp. 1–12 (2009)
36. Kovachev, D., Renzel, D., Klamma, R., Cao, Y.: Mobile community cloud computing: Emerges and evolves. In: 11th International Conference on Mobile Data Management (MDM'10)., pp. 393–395 (2010)
37. Kwok, Y.K., Song, S., Hwang, K.: Selfish grid computing: game-theoretic modeling and NAS performance results. In: CCGRID, pp. 1143–1150 (2005)
38. Litzkow, M.J., Livny, M., Mutka, M.W.: Condor - A hunter of idle workstations. In: ICDCS '88: 8th International Conference on Distributed Computing Systems, pp. 104–111 (1988)
39. Qian, H., Medhi, D.: Server operational cost optimization for cloud computing service providers over a time horizon. In: Proceedings of the 11th USENIX conference on Hot topics in management of internet, cloud, and enterprise networks and services (Hot-ICE'11) (2011)
40. Ranganathan, K., Ripeanu, M., Sarin, A., Foster, I.T.: Incentive mechanisms for large collaborative resource sharing. In: CCGRID, pp. 1–8 (2004)
41. Ranjan, R., Rahman, M., Buyya, R.: A decentralized and cooperative workflow scheduling algorithm. Cluster Computing and the Grid, IEEE International Symposium on $0$, 1–8 (2008)
42. Rappa, M.A.: The utility business model and the future of computing services. IBM Syst. J. $43$, 32–42 (2004)
43. Ross, J.W., Westerman, G.: Preparing for utility computing: The role of it architecture and relationship management. IBM Systems Journal $43$(1), 5–19 (2004)
44. Rzadca, K., Trystram, D., Wierzbicki, A.: Fair game-theoretic resource management in dedicated grids. In: Proceedings of the Seventh IEEE International Symposium on Cluster Computing and the Grid (CCGRID 07), pp. 343–350 (2007)
45. Schulz, S., Blochinger, W., Held, M., Dangelmayr, C.: Cohesion - a microkernel based desktop grid platform for irregular task-parallel applications. Future Generation Computer Systems $24$(5), 354–370 (2008)
46. SpotCloud. http://www.spotcloud.com/

47. Stokely, M., Winget, J., Keyes, E., Grimes, C., Yolken, B.: Using a market economy to provision compute resources across planet-wide clusters. In: Proceedings of the 2009 IEEE International Symposium on Parallel&Distributed Processing (IPDPS 2009), pp. 1–8 (2009)
48. Vecchiola, C., Pandey, S., Buyya, R.: High-performance cloud computing: A view of scientific applications. In: Proceedings of the 2009 10th International Symposium on Pervasive Systems, Algorithms, and Networks (ISPAN '09), pp. 4–16 (2009)
49. Velte, T., Velte, A., Elsenpeter, R.: Cloud Computing, A Practical Approach, 1 edn. McGraw-Hill, Inc., New York, NY, USA (2010)
50. Weinman, J.: Cloudonomics: The Business Value of Cloud Computing. Wiley (2012)
51. Wolski, R., Plank, J.S., Brevik, J., Bryan, T.: G-commerce: Market formulations controlling resource allocation on the computational grid. In: Proceedings of the 15th International Parallel & Distributed Processing Symposium (IPDPS 2001), pp. 46–53 (2001)
52. Xu, K., Song, M., Zhang, X., Song, J.: A cloud computing platform based on p2p. In: IEEE International Symposium on IT in Medicine Education (ITIME'09)., vol. 1, pp. 427–432 (2009)
53. Zaman, S., Grosu, D.: Combinatorial auction-based allocation of virtual machine instances in clouds. In: IEEE CloudCom, pp. 127–134 (2010)
54. Zhang, Q., Cheng, L., Boutaba, R.: Cloud computing: state-of-the-art and research challenges. Journal of Internet Services and Applications 1(1), 7–18 (2010)
55. Zhao, H., Li, X.: Efficient grid task-bundle allocation using bargaining based self-adaptive auction. In: Proceedings of the 9th IEEE/ACM International Symposium on Cluster Computing and the Grid (CCGrid 2009), vol. 0, pp. 4–11 (2009)
56. Zhao, H., Liu, X., Li, X.: Hypergraph-based task-bundle scheduling towards efficiency and fairness in heterogeneous distributed systems. In: Proceedings of the 24th IEEE International Parallel and Distributed Processing Symposium (IPDPS'12), pp. 1–12 (2010)
57. Zhao, H., Liu, X., Li, X.: A taxonomy of peer-to-peer desktop grid paradigms. Cluster Computing 14(2), 129–144 (2011)
58. Zhao, H., Pan, M., Liu, X., Li, X., Fang, Y.: Optimal resource rental planning for elastic applications in cloud market. In: Proceedings of the 26th IEEE International Parallel and Distributed Processing Symposium (IPDPS'12) (2012)
59. Zhao, H., Yu, Z., Tiwari, S., Mao, X., Lee, K., Wolinsky, D., Li, X., Figueiredo, R.: Cloudbay: Enabling an online resource market place for open clouds. In: IEEE Fifth International Conference on Utility and Cloud Computing (UCC'12), pp. 135–142 (2012)

# Chapter 2
# Optimal Resource Rental Management

**Abstract** Application services using cloud computing infrastructure are prolifer-
ating over the Internet. In this chapter, we study the problem of how to minimize
resource rental cost associated with hosting such cloud-based application services,
while meeting the projected service demand. This problem arises when applications
incur significant storage and network transfer cost for data. Therefore, an Application Service Provider (ASP) needs to carefully evaluate various resource rental
options before finalizing the application deployment. We choose Amazon® EC2
marketplace as a case of study, and analyze the optimal strategy that exploits the
tradeoff of data caching versus computing on demand for resource rental planning in
cloud. Given fixed resource pricing, we first develop a deterministic model, using a
mixed integer linear program, to facilitate resource rental decision making. Next, we
investigate planning solutions to a resource market featuring time-varying pricing.
We conduct time-series analysis over the spot price trace and examine its predictability using Auto-Regressive Integrated Moving-Average (ARIMA). We also
develop a stochastic planning model based on multistage recourse. By comparing
these two approaches, we discover that spot price forecasting does not provide our
planning model with a crystal ball due to the weak correlation of past and future
price, and the stochastic planning model better hedges against resource pricing
uncertainty than resource rental planning using forecast prices.

This chapter is organized as follows. Section 2.1 provides an overview of the
problem and summarizes our proposed optimal planning methods. Section 2.2
surveys the related work. In Sect. 2.3, we formulate the system model, provide
a deterministic planning model for the resource rental problem, and evaluate the
performance of the deterministic pricing resource planning approach. Finally, in
Sect. 2.4, we analyze the predictability of Amazon® EC2 spot pricing using time-
series analysis techniques, propose a stochastic optimization model to solve the
rental planning problem, and perform simulations to compare the two approaches.

H. Zhao and X. Li, *Resource Management in Utility and Cloud Computing*,
SpringerBriefs in Computer Science, DOI 10.1007/978-1-4614-8970-2_2,
© The Author(s) 2013

## 2.1  Overview

The emerging cloud computing model, with its virtually infinite resources and elasticity, liberates organizations from the expensive infrastructure investment. As a result, more and more Application Service Providers (ASPs) recognize the separation between the actual application and the infrastructure necessary to run it, and begin to deploy applications on resources rented from infrastructure providers. According to a recent forecast by Gartner® [16], Software-as-a-Service and Cloud-based business application services will grow from $13.4 billion in 2011 to $32.2 billion in 2016.

In cloud computing, a major issue faced by the ASPs is how to minimize the resource rental cost while meeting their application service demand. Significant research efforts have been directed toward developing optimal resource provisioning schemes to meet service requirements (avoid the cost due to over-provisioning and the penalty due to under-provisioning) [7, 17, 23, 28, 30]. These works, although offer effective resource provisioning controls in response with varying workload, are still coarse-grained in terms of exploring application elasticity with regard to different resource pricing options. We believe that resource rental planning should be conducted in a cost-aware manner to reduce ASPs' operational cost. Specifically, we propose a fine-grained planning scheme to regulate the rental activities on a time-slotted basis, exploring hourly charging rate of various types of resources, in order to meet the projected service demand and minimize resource rental cost at the same time. Complementary to prior resource scaling solutions, our approach focuses on application scaling that optimizes resource rental plan in cloud without compromising the service-level agreement.

In addition to the planning optimization complexity, another obstacle lies in the uncertainty of computational resource pricing. This challenge is encountered in the spot resource market emerged in recent years. In a spot resource market, depending on the resource supply and demand level, the unit price of a computational instance is fluctuating all the time. For example, at the time Amazon® first launched its spot instance service in December 2009, an auction mechanism was employed to determine instance pricing. Since spot instances leverage idle cycles from the regular on-demand server pool, they are auctioned off at a price much lower than that of the regular on-demand instances most of the time. As a result, this real-time bidding market has attracted many ASPs who wish to increase server capacity at low cost. There is a growing research interest in utilizing spot instance service. However, modeling and analyzing spot instance pricing is largely neglected due to the lack of demand and resource provision information. We believe that our study is helpful to understand spot pricing, and more importantly, to improve resource utilization under spot pricing.

The research presented in this chapter represents our initial design for cost-effective resource utilization and management in utility and cloud computing. In particular, we develop optimal resource rental planning strategies for fixed pricing and stochastic pricing resource markets, respectively. The first part of this chapter presents our approach for a fixed pricing resource market. Given a forecast demand schedule, the ASP needs to periodically review the running progress of the deployed service and make optimal job allocation as well as resource rental decisions so as not to waste money on excessive computation, storage or data transfer. We formulate a deterministic planning model for resource rental decision making over a specific planning horizon. The solution to this model serves as a guide to make cost-effective resource rental decisions in real time. We show that our planning model is especially useful for high-cost Virtual Machine (VM) classes. This is because cost saving from our model primarily comes from eliminating unnecessary job running by decreasing VM rental frequency. From this perspective, our model formulation is aligned with the dynamic lot-sizing model commonly encountered in the field of production planning.

Next, we analyze and solve the fine-grained cloud resource rental planning problem under the pricing uncertainty challenge. In particular, two possible solutions are jointly explored. We systematically analyze the predictability of Amazon® EC2 spot pricing and use the predictive prices to perform planning. Furthermore, we propose a multistage resource model for stochastic resource rental planning. This model decomposes the stochastic process of decision making under varying price into sequential decision making processes with the aid of price distribution at various stages. As such, the stochastic optimization problem is transformed into a large-scale deterministic optimization problem. Through simulations, we demonstrate that the stochastic planning approach is more cost-effective than predictive planning.

## 2.2 Related Work

Nowadays, a wide variety of computational and data intensive applications utilize cloud to their benefit. Therefore, it becomes imperative to understand the cost-benefit of running resource-demanding applications in cloud in order to make cost-effective resource rental decisions. Cloud computing eliminates up-front setup and operational cost for distributed resources. However, moving and storing large data set in cloud incur significant cost comparable to the computing cost [13]. Efforts have been made to mitigate such cost in cloud [22, 29]. In this chapter, we present a planning model that optimizes resource usages for elastic applications with comprehensive cost considerations.

Finding an optimal resource utilization strategy is challenging for both cloud infrastructure providers and application service providers who rely on rented infrastructure. From the perspective of the cloud infrastructure provider, the challenge is how to reduce the operational cost and maximize leasing revenue. Many existing research has focused on this aspect. The general problem of minimizing resource allocation cost while meeting job demand is NP-hard [9]. Resource scheduling for the emerging spot market was proposed in [31]. The proposed framework includes: (1) a market analyzer periodically forecasting supply and demand, (2) a capacity planner determining the spot price based on the forecast results, and (3) a VM scheduler maximizing the revenue by solving a NLIP model for the scheduling problem. From the perspective of the application service providers, the challenge becomes how to minimize resource rental cost while meeting service demand from customers. Many resource planning schemes rely on predictive workload assessment [17, 21]. Our work takes one step further that presents an application scaling control model based on the forecast demand. Our model takes full consideration of various resource types and their associated costs within a cloud resource market, and strives to find the optimal tradeoff among various resource usage in resource rental allocation.

The stochastic planning model proposed in this chapter deals with the price uncertainty in the spot resource market. Such a spot market is either formed by multiple resource providers [10] or by a single resource provider. Amazon® EC2 spot market is the most representative example that attracts significant research attentions. Researchers are interested in utilizing spot instances to temporarily add capacity to dedicated clusters during peak periods [19]. The biggest concern for utilizing spot instances is that it is hard to guarantee resource availability. Recent works [3, 20] addressed this problem using statistical analysis. Notably, Ben-Yehuda et al. [1] reversely engineered spot prices by constructing a spare capacity pricing model based on existing price traces. However, the effectiveness of these approaches is still unclear due to unsubstantiated assumptions on Amazon® EC2 spot service. In this chapter, we take EC2 as a case study and targets at a general spot resource market where prices are market-driven and users bid according to their true valuations (simple-minded assumption). The most relevant works to this study are presented in [8, 25]. In [8], the authors presented an optimal VM placement algorithm that minimizes the cost of resource provisioning in a multiple cloud providers environment, and in [25], the authors proposed a profit-aware dynamic bidding algorithm to optimize ASP's profits in EC2 spot market. Our work's application scenario is different from [8], and we develop our model based on realistic application and price traces. Comparing with [25], our approach proposes a different model that takes storage and network transfer cost into account in addition to computational instance bidding.

## 2.3 Deterministic Resource Rental Planning

Resource rental planning entails the acquisition and allocation of computational and storage resources to applications so as to satisfy demand over a specified time horizon. An application scaling control scheme is proposed to optimize rental decision on a time slotted basis. In this section, we target at a fixed pricing cloud resource market. After describing the system model, we model the rental planning problem using a mixed integer linear program.

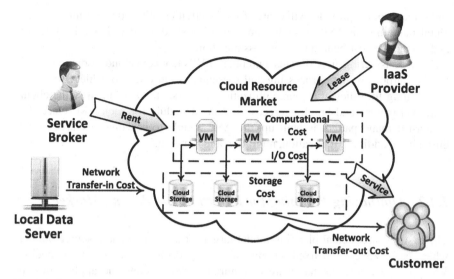

**Fig. 2.1** System model for the resource rental planning problem

## 2.3.1   System Model

We present a scenario where an ASP offers some computational and data intensive application services (example services are data visualization, data analytics, data indexing, etc.) to customers over a network. Instead of using local resources, the tasks of computation and data storage are completely outsourced to a shared resource pool operated by some Infrastructure-as-a-Service (IaaS) provider(s), shown in Fig. 2.1. The depicted system model resembles a broad range of practical examples in today's cloud-based service market. For instance, the ASP could be mapped to some Software-as-a-Service provider who offers routine data analytics to its customer firms, or some academic institution that provides scientific data visualization services to the general public.

As illustrated in Fig. 2.1, resource usage incurs monetary cost to the ASP in various forms. Rental activities are charged throughout the life cycle of the deployed service as follows. First, input data is imported into the cloud from the local storage media, introducing network transfer-in cost. Next, a number of VM instances (hereby referred as Virtual Servers, or VS for short) are launched to perform data processing tasks. Each of them costs certain amount of money depending on both VS unit price and rental duration. After the computational jobs are completed, results and logs are saved to cloud storage, and may later be dumped into local persistent storage. Many often the data size is large (e.g., images or videos) and incurs significant storage and network transfer-out cost for the ASP. The storage cost may also apply to input data already fetched into the cloud but not processed yet. Finally, high performance applications often feature tremendous I/O requirements and some resource provider will charge for I/O activities. When performing resource rental planning, an ASP needs to consider all costs described above in order to understand the cost-benefit ratio of possible choices.

Now, considering an ASP rents a number of VSs from the cloud resource market for the purpose of data processing and presentation, in order to achieve resource auto-scaling for efficient resource utilization, the first step is to identify the client workload pattern and build a forecast demand schedule for each VM. Once the forecast demand pattern is built up, the ASP is able to schedule resource rental through job addition, replication, migration and removal.

### 2.3.2  Optimizing Planning for Deterministic Pricing Market

The first resource rental planning model targets at an on-demand resource market where each VS costs a fixed amount of money. Each VS belongs to a specific VS type specifying the hardware configuration. We assume the applications to be elastic and composed of jobs easy to scale gracefully and automatically. For example, applications processing Bags-of-Tasks (no job dependencies). Similar to [14], we are interested in self-aware solutions that can plan resource usage of cloud applications under various pricing. The planning horizon $T$ is divided into fixed time slots $t = 1, \ldots, T$. We refer the start of each time slot as a *decision point*. At each decision point, a rental operation is performed to access the most cost-effective resource available for the application.

Let $\mathscr{T}$ be the set of decision points. The goal of resource rental planning is to minimize the total rental cost associated with processing the forecast workload over the planning horizon $T$. In order to accomplish this goal, three sets of variables are introduced to identify the rental decisions to be made at each decision point. The first set of variables, $\alpha_{i,t}$, denotes the amount of data to be processed by the application during time slot $t$ on a type-$i$ VS. Next, at the end of slot $t$, we use $\beta_{i,t}$ to represent the desired storage space for holding the data. Finally, let binary decision variables $\chi_t$ denote if powering on a type-$i$ VS is needed at time slot $t$. $\alpha_{i,t}$ and $\chi_{i,t}$ specify how to make use of the computational resources to control the application progress, while $\beta_{i,t}$ determines the amount of storage resources to reserve in a cloud market. If all these variables are determined, an application scaling control policy is formed to guide the rental activities in the cloud market for optimal resource utilization.

A number of cost parameters are associated with our resource rental optimization problem. Specifically, the rental cost (processing cost) for type-$i$ VS in time slot $t$ is $C_p(i,t)$, and the storage rental cost per data unit for slot $t$ is $C_s(t)$. As presented earlier in Sect. 2.3.1, many IaaS providers charge nontrivial cost for data transfer across the cloud boundary. For each time slot $t$, let $C_{io}(t)$ be the I/O cost for data transfer from and to the cloud storage, and let $C_f^+(t)$ and $C_f^-$ be the cost for transferring into and out of the cloud, respectively. In addition to the cost parameters, we assume the customer's demand function is $D(\cdot)$, where $D(i,t)$ denotes the forecast workload demand profile for a type-$i$ VS in slot $t$. We summarize the notation used throughout the chapter in Table 2.1.

**Table 2.1** Summary of notations

| Variables | |
|---|---|
| $\alpha_{i,t}$ | Output data size generated by one type-$i$ VS in time slot $t$ |
| $\beta_{i,t}$ | Storage space for data produced by one type-$i$ VS at the end of slot $t$ |
| $\chi_{i,t}$ | Binary decision variable representing rental decision of one type-$i$ VS in time slot $t$ |
| **Parameters** | |
| $\mathscr{T}$ | Set of time slots |
| $\mathscr{I}$ | Set of VS types |
| $C_p(i,t)$ | VS rental cost (per type-$i$ VS · slot duration) |
| $C_s(t)$ | Storage cost (per data unit · slot duration) |
| $C_{io}(t)$ | I/O operation cost (per data unit · slot duration) |
| $C_f^+(t)$ | Network transfer-in cost (per data unit · slot duration) |
| $C_f^-(t)$ | Network transfer-out cost (per data unit · slot duration) |
| $D(i,t)$ | Demand to be satisfied for one type-$i$ VS at the end of slot $t$ |
| $P(i)$ | Average bottleneck resource consumption rate (per data unit generated) for one type-$i$ VS |
| $Q(i,t)$ | Bottleneck resource available for one type-$i$ VS in time slot $t$ |
| $\Phi_i$ | Average output-to-input ratio for one type-$i$ VS (application specific) |

With all the prerequisites, we formulate the rental payment function following a linear cost model. More specifically, the rental cost is linearly proportional to the consumed resource amount as well as to the duration of the rental period. Naturally, our objective function aims at minimizing the rental cost for each type-$i$ VS over the entire planning horizon $T$. At each decision point, a fixed rental cost $C_p(i,t)$ is charged if the ASP decides to rent one type-$i$ VS ($\chi_{i,t} = 1$). Now, given the presence of this computational resource cost, the ASP may choose to make full use of the VS capacity so as to meet the forecast workload demand over a number of future time slots. However, doing so will increase the storage and I/O cost as more workload is processed earlier in time. As such, the planning problem emerges as the ASP needs to carefully trade off the computational rental cost versus storage and data migration costs. In production planning, similar problems are recognized as the dynamic lot-sizing problem. The solution to the dynamic lot-sizing problem determines the optimal frequency of setups so as to minimize the total cost within the resource and demand constraints. In the context of cloud computing, we formulate the planning problem under fixed resource pricing as the **D**eterministic **R**esource **R**ental **P**lanning (DRRP) problem. DRRP models cloud resource rental on a per-VS basis, forming a fine-grained control policy for rental planning. The complete model formulation is given as follows.

$$\min \sum_{t \in \mathcal{T}} (C_f^+(t) \cdot \Phi_i \cdot \alpha_{i,t} + (C_s(t) + C_{io}(t))$$

$$\cdot \beta_{i,t} + C_f^-(t) \cdot D(i,t) + C_p(i,t) \cdot \chi_{i,t}) \tag{2.1}$$

s.t.

$$\beta_{i,t-1} + \alpha_{i,t} - \beta_{i,t} = D(i,t), \qquad i \in \mathcal{I}, t \in \mathcal{T} \tag{2.2}$$

$$P(i) \cdot \alpha_{i,t} \leq Q(i,t), \qquad i \in \mathcal{I}, t \in \mathcal{T} \tag{2.3}$$

$$\alpha_{i,t} \leq B \cdot \chi_{i,t}, \qquad i \in \mathcal{I}, t \in \mathcal{T} \tag{2.4}$$

$$\beta_{i,0} = \varepsilon, \qquad i \in \mathcal{I} \tag{2.5}$$

$$\alpha_{i,t}, \beta_{i,t} \in \mathbb{R}_+, \qquad i \in \mathcal{I}, t \in \mathcal{T} \tag{2.6}$$

$$\chi_{i,t} \in \{0,1\}, \qquad i \in \mathcal{I}, t \in \mathcal{T} \tag{2.7}$$

Note that the objective function does not take I/O and storage cost for input data into account. This is because we assume that input data is brought into cloud on the fly to complete the computational jobs. Another option is to copy all input data once and store them in cloud throughout the entire planning horizon. The decision on which option is better depends on the data access pattern and the duration of planning horizon. Here, we assume that input data is "transfer-on-demand" to simplify the presentation.

Constraint (2.2) is analogous to the inventory balance constraint in the dynamic lot-sizing problem. It specifies that workload demand should be met at any time slot. At slot $t$, the data stored at the previous time slot $\beta_{i,t-1}$, and the data generated in the current slot $\alpha_{i,t}$, are combined together to serve the forecast demand profile emerged in the current time slot, i.e., $\beta_{i,t-1} + \alpha_{i,t} \geq D(i,t)$. The overprovisioning amount becomes the storage amount $\beta_{i,t}$ at the end of $t$. The initial storage space is set to be some constant $\varepsilon$ in constraint (2.5), depending on the specific planning scenario. Next, let $P(i)$ be the average bottleneck resource consumption rate for one type-$i$ VS, and let $Q(i,t)$ denote the bottleneck resource available for one type-$i$ VS in $t$, constraint (2.3) ensures that the workload processing rate does not saturate the available bottleneck resource.

Constraint (2.4) is often referred to as the forcing constraint. It states that there will be no data generated in $t$ if no rental decision is made ($\chi_{i,t} = 0$). $B$ is set to be a very large constant that exceeds the maximum possible value of $\alpha_{i,t}$. Finally, constraints (2.6) and (2.7) specify domains of the variables.

The formulation of DRRP is a mixed integer linear program (MILP) that is NP-complete in nature. With reasonable input size, this problem can be solved using standard techniques such as the branch-and-bound (B&B) method. These algorithms are implemented in many optimization software packages. For more details with regard to the algorithms, we refer readers to [32].

### 2.3.3 Evaluation of DRRP

We consider three VS classes $\mathscr{I} = \{$c1.medium, m1. large, m1.xlarge$\}$, and perform simulations to evaluate the solution to DRRP based on realistic pricing and application-usage scenarios. The rental planning decisions are calculated in an hourly basis, spanning over daily planning horizon (24 h). The MILP formulation is solved by the CPLEX$^{TM}$ [12] solver integrated in AIMMS 3.11 [2]. We sample the hourly data processing demand from a normal distribution $\mathscr{N}(0.4, 0.2)$ (expressed in the unit of Gigabyte). It is assumed that the software required by the application services has been configured on virtual servers rented from the cloud market. Therefore, we do not take the initial environment preparation into account.

The cost parameters used in model formulations are set according to Amazon®'s EC2 on-demand pricing policy.[1] Specifically, the hourly on-demand VS rental costs are {$0.2, $0.4, $0.8} for the three VS classes. Using Elastic Block Store (EBS), the storage cost is $0.1 per GB/month, and 0.1 per million I/O operations. The inbound and outbound transfer cost is $0.1 and $0.17 per GB. In order to provide realistic parameter estimates in our proposed models, we refer to a recent paper [4] studying the cost and performance of running scientific workflow applications on Amazon® EC2. Based on the 3-year cost of a mosaic service (generated by an astronomical application Montage, see [18] for details) hosted on EC2, we normalize the I/O cost to $0.2 per GB, and set $\Phi_i$ to 0.5 for all $i \in \mathscr{I}$. According to the data provided in [4] (runtime, input and output volume, etc.), the virtual servers are able to offer sufficient resources for serving the randomly generated demand. Therefore, constraint (2.3) in DRRP is omitted.

We first show the cost-saving advantage of our proposed solution over resource rental without planning. The results are shown at the upper side of Fig. 2.2. In our simulation, per-VS costs over daily planning horizon for both schemes are compared. From the results, we observe that cost derived from solving DRRP is significantly lower than that of the no-planning solution. As VS becomes more powerful, the cost reduction becomes more significant. Especially, the cost reduction for VS of class m1.xlarge achieves nearly 50% drop off. This is because compared to the no-planning solution, the cost reduction primarily comes from the saving of computational cost (virtual servers are turned off in cloud when demand is satisfied by cached data in cloud storage). Therefore, more saving is expected for high-cost VS classes. The cost structure for each VS class is presented in the lower side of Fig. 2.2. The proportion of computational cost is relatively stable in all three classes. However, we observe that more money is spent on I/O and storage as VS becomes more powerful. This is because more powerful VS incurs higher VS rental cost each time the rental decision is made. As a result, an ASP tends to utilize caching more often to serve the customer demand and rents VS less frequently.

---

[1]Amazon® has declared lower pricing for EC2 when we prepared this manuscript. Since our simulation is based on [4], the study presented here is by no means up-to-date, but serves as a representative case of study.

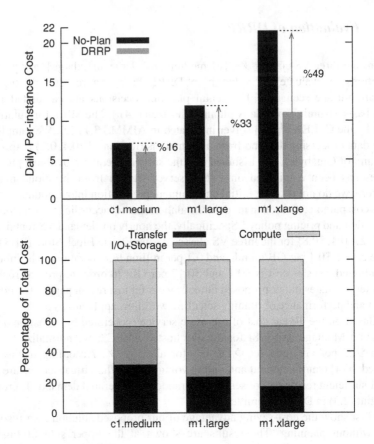

**Fig. 2.2** Cost analysis of DRRP

Next, we conduct a sensitivity analysis to the solution for DRRP and plot the results in Fig. 2.3. We define cost ratio as the cost of rental planning based on DRRP to the cost of resource rental without planning. The base ratio (67%) is set to the cost ratio of VS class m1.large calculated in the last simulation. From this base ratio, we first vary the weights of I/O and computational cost gradually. In one direction, we keep the I/O cost fixed and increase the computational cost with a fixed step of 0.1, and then we increase the I/O cost in the other direction similarly. The result showed in the left part of Fig. 2.3 clearly demonstrate that the cost reduction achieved by solving DRRP becomes more salient for expensive computational resources. This conclusion confirms the analysis we previously provided. The impact of demand is investigated in the right part of Fig. 2.3. In particular, we alter the mean of the demand distribution from 0.2 to 1.6 GB/h. As more demand is generated for services, the computational resources tend to be kept busy all the time because the current storage cannot meet the demand. As a result, cost reduction is not noticeable for heavy service demand.

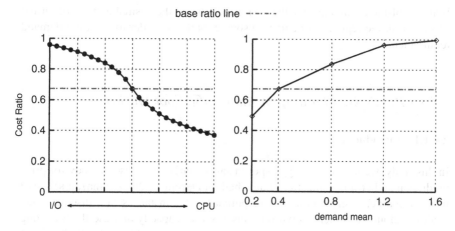

**Fig. 2.3** Sensitivity analysis for DRRP

## 2.4 Dealing with Spot Pricing Uncertainty in Cloud

In this section, we extend the resource rental planning model by including cost uncertainty. Such uncertainty is introduced by many IaaS providers who offer a spot pricing option for idle computational resources. Example markets can be found in [15, 26]. The price fluctuation of spot resources over time creates time series data for analysis. Using Amazon®'s spot market as a case of study, we take two routes to attack the resource rental planning problem with spot pricing uncertainty. First, we apply time series forecasting to spot price history crawled from [11]. The prediction results are then fed into our deterministic planning model (hereafter labeled as **predictive planning**). Next, we propose an alternative approach that leverages the price distribution information (hereafter labeled as **stochastic planning**). A dynamic programming algorithm is also presented to solve the stochastic optimization problem. We compare the two approaches in the end of this section.

Before we proceed, a few assumptions need to be clarified. First, we assume that ASPs will bid truthfully in the spot resource acquisition process. This assumption is in line with the assumption made in [20]. With this assumption, an ASP will not bid strategically. In fact, whether strategic bidding is helpful to achieve some desired level of resource availability is controversial. On the one hand, by exploiting prior price history, it is viable to optimize bidding using probabilistic models for a single bidder [3]. On the other hand, one should also consider bidding strategies of other bidders before making decisions. From a game theoretic perspective, intentionally overbidding or underbidding is not a dominant strategy (e.g., if every bidder overbids, the spot price increases, only benefiting the IaaS provider). Second, an *out-of-bid* event occurs when an ASP's bid price is lower than the spot price.

If an out-of-bid event happens, the ASP needs to rent the desired number of virtual servers from the regular on-demand resource market in order to meet the demand requirement.

## 2.4.1  Predictive Planning in Amazon® Spot Market

### 2.4.1.1  Introduction

In this study, we use Amazon®'s spot instance market as a case of study for price prediction and cost optimization. Launched on December 2009, Amazon®'s spot instance market offers a new way to purchase EC2 instances in a discount rate. It allows cloud customers to bid on unused server capacity and use them as long as the bid exceeds the current spot price, which is updated periodically based on supply and demand. Payment in spot instance auction is uniform, i.e., all winners in the auction will pay a per-unit price equal to the lowest winning bid (a.k.a the spot price). While running spot instances saves huge cost (typically over 60 % according to [27]), it also introduces significant uncertainty for resource availability. As a result, previous resource rental planning model based on deterministic resource pricing does not apply.

If one is able to forecast spot prices with relatively high accuracy, then these predictions can be used to instantiate the DRRP model presented in Sect. 2.3.2 to obtain a near-optimal solution. However, performing forecasting is challenging for customers because they do not possess the global information of supply and demand as Amazon® does. In [31], the authors attempted to predict customer demand from the view of an IaaS provider. They proposed a simple auto-regression model for prediction but no prediction results were reported due to the lack of realistic demand information. Another study on the predictability of Amazon®'s spot instance price was presented in [20]. Their work focused on achieving availability guarantee with spot instances, and used a quantile function of the approximate normal distribution to predict when the autocorrelation of current and past price is weak. When the autocorrelation is strong, a simple linear regression prediction model was adopted. However, we found that such an approximation is inaccurate in some test cases that cannot be taken as a generic approach. In this section, we will assess the predictability of spot instance price based on a statistical approach (ARIMA), and estimate the prediction errors using empirical data set.

### 2.4.1.2  Methodology

We collected the historical data (published in [11]) for spot price variation from February 1, 2010 to June 22, 2011. The data source represents spot price variations for Linux instances in *us-east-1* region. The data size is approximately 100K records. We employ a statistical approach to analyze the predictability of Amazon® EC2 spot pricing, and plot the results in Fig. 2.4.

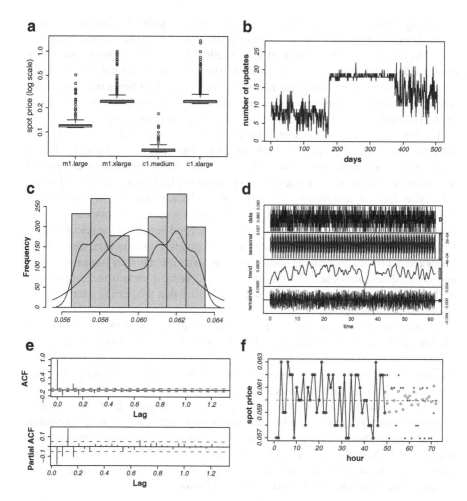

**Fig. 2.4** Analyzing the predictability of Amazon® EC2 spot price. (**a**) Box-and-Whisker diagram; (**b**) frequency analysis; (**c**) histogram plot; (**d**) price decomposition; (**e**) correlation analysis; (**f**) 24-h prediction

The first step in our investigation is to identify the outliers in the original data set. Figure 2.4a plots the box-and-whisker diagram for the spot price data set corresponding to four different Linux VM classes. The outliers are identified as those points beyond the whiskers (1.5 IQR (interquartile range) of the upper quartile). We can see that more outliers present in more powerful VM class, indicating increasing price dynamics in more powerful types. However, even for the most powerful instance (c1.xlarge), the number of outliers still contributes a trivial amount to the overall data set (<3 %).

Having trimmed out the outliers, we still cannot apply standard time series analysis because the derived data set is unequally spaced with inconsistent sampling intervals, as shown in Fig. 2.4b. It plots the daily price update frequency for VS of class *linux-c1-medium*. For that reason, we further convert the data into equally spaced time series data with a regular update frequency of 24 times per day. At the start of each hour, the spot price is set to be the most recent updated price in the last hour. If no update appears in the last hour, the spot price is unchanged.

We have performed various experiments on this converted data set, each with different time scale of prediction (both short-term and long-term). Here we show a representative prediction result for instance of class *linux-c1-medium* over a period of 2 months. Specifically, we use the data ranging in [12/1/2010, 1/31/2011] as the estimation data set, and data in 2/1/2011 as the validation data set. In other words, the data collected from the 2-month historical records is used to provide the next-day price forecasting. In Fig. 2.4c, we plot the histogram and density of the selected data. We also randomly generate the same number of points from a normal distribution characterized by the three main measures in quantitative statistics (mean, variance and standard deviation), and plot the curve in Fig. 2.4c for comparison. Examination of the Shapiro-Wilk test result (omitted here) verifies that the pricing data does not fit the normal distribution.

In order to identify patterns in the selected series and perform prediction, we use the ARIMA approach developed by Box and Jenkins [6], which retains great flexibility in recognizing data patterns and is relatively lightweight compared to machine learning techniques such as artificial neural networks or support vector machines. Two common processes are used in ARIMA to identify the correct time series pattern. The first process is the Auto-Regressive (AR) process that decomposes observations into a random error component and a linear combination of prior observations. The second process is called the Moving Average (MA) process. In MA, each observation is made up of a random error component, and a linear combination of prior random errors. Given a time series of data $X_t$, the general form of an ARIMA process is given as follows:

$$\left(1 - \sum_{i=1}^{p} \phi_i L^i\right)(1 - L)^d X_t = \left(1 + \sum_{i=1}^{q} \theta_i L^i\right)\varepsilon_t, \qquad (2.8)$$

where L is the lag operator, $\phi_i$ and $\theta_i$ are the parameters for AR and MA process, respectively, and $\varepsilon_t$ are error terms. The key to the ARIMA model is to identify parameters $p$ (AR parameter), $d$ (differencing pass), and $q$ (MA parameter) correctly. This is achieved through a series of steps. First, we verify that our test data series is statistically stationary (statistical properties such as mean and variance are constant over time), and does not require further differencing. The decomposition of the selected series is presented in Fig. 2.4d, where the original time series is decomposed into three parts: trend, seasonal, and random noise. We can see that the target series does not exhibit clear trend, but advertises certain cyclic pattern as shown in the seasonal decomposition. For that reason, we revise our prediction approach by employing a Seasonal ARIMA (SARIMA) model, which takes the seasonal component into account. It can be expressed as SARIMA, $(p, d, q) \times (P, D, Q)_{24}$, which includes the seasonal parameters for price prediction.

The next step for identifying the SARIMA model parameters is to plot the correlograms for autocorrelation function (ACF) and partial autocorrelation function (PACF), as displayed in Fig. 2.4e. These two functions help to detect trend and seasonality of the selected series. Note that the x-axis is normalized by frequency so that 1.0 corresponds to lag = 24. From the graphs we observe that, the selected series has certain degree of correlation with its past at certain lag value, e.g., lag = 3, because these values exceed the 95 % confidence limit. However, such a correlation is not strong enough since its value is still far from 1.0 (which indicates perfect correlation).

Finally, the identification of the most appropriate model parameters is achieved by the forecast package developed in R [24]. In the forecast package, the calling of *auto.arima* function will return the best model according to Akaike information criterion (AIC) or Bayesian information criterion (BIC) values. The function performs a search over possible models within the order constraints provided. Through extensive trials, we found that most test series fit SARIMA $(2, 0, 1 \text{ or } 2) \times (2, 0, 0)_{24}$ best. The prediction result for the selected series is shown in Fig. 2.4f. The solid points and the hollow points represent the predicted and the actual prices on February 1st, 2011, respectively. The horizontal dashed line represents the average price in the selected data series, while the fluctuating solid lines represent spot price variation in the past 48 h. We observe that the predicted prices are mostly hanging over the average price line. While this model returns the least prediction error compared to other models, its mean squared prediction error (MSPE) is only slightly better than the simple prediction using the expected mean value.

## 2.4.2 Stochastic Planning for Spot Pricing Market

### 2.4.2.1 Solution Overview

In addition to the predictive planning approach, we propose an alternative approach that takes the stochastic nature of the spot pricing into account. We model the fluctuation of the spot instance rental cost $C_p(i, t)$ as a stochastic process $C_p$ with state space $\mathbb{S}$. $C_p$ is a collection of $\mathbb{S}$-valued random variables on a probability space $\Omega$ indexed by the time slot set $\mathcal{T}$, i.e., $C_p$ for class-$i$ instance is a collection: $\{C_p(i, t) : t \in \mathcal{T}\}$. The true valuations of the spot prices over the planning horizon are represented by set: $\{\widehat{C_p}(i, t) : t \in \mathcal{T}\}$. The goal of the stochastic resource rental planning is to optimize the expected overall cost over the complete state and probability space. In particular, the objective function (2.1) in DRRP can be reformulated as follows:

$$\delta_{exp} = \mathbf{E}_{C_p} \left\{ \sum_{t \in \mathcal{T}} (C_f^+(t) \cdot \Phi_i \cdot \alpha_{i,t} + (C_s(t) + C_{io}(t)) \right.$$

$$\left. \cdot \beta_{i,t} + C_f^-(t) \cdot \mathrm{D}(i,t) + C_p(i,t) \cdot \chi_{i,t}) \right\}, \tag{2.9}$$

where $\delta_{exp}$ is the expected total cost. The optimization model now becomes to minimize (2.9), subject to constraints (2.2)–(2.7). We summarize our solution to stochastic resource planning as follows.

1. Generate bid prices $\widehat{C_p}(i,t)$ for the class-$i$ VS at every $t \in \mathcal{T}$, based on the true valuations.
2. Calculate the base probability distribution according to the pricing history.
3. Derive new probability distributions at all $t \in \mathcal{T}$ according to the base distribution and the bid price.
4. Reformulate using a multistage recourse approach, based on the newly generated distributions.
5. Solve the deterministic equivalent reformulation.

Due to the possibility of losing the auction, the actual realizations of spot prices are possibly different at multiple decision points. Steps (1)–(3) summarize our solution to this challenge. We call our proposed approach bid-dependent dynamic sampling. After calculating the distributions, a multistage resource model is used to optimize the expected total cost.

### 2.4.2.2  Bid-Dependent Dynamic Sampling

Let $\mathbb{S}_i$ be the finite state space for the spot price of a class-$i$ VS. A base probability distribution is the summarized discrete probability distribution over a selected historical price series: $Pr(C_p(i,t) = s_i), s_i \in \mathbb{S}_i$. This distribution cannot be used in our stochastic optimization model because it does not include the risk of out-of-bid. Therefore, we propose to use the following approach to dynamically generate the probability distribution at every decision point $t$. The values in the finite state space $\mathbb{S}_i$ is sorted in the ascending order (no equivalent values are present in $\mathbb{S}_i$). Suppose the fixed on-demand cost is $\lambda_i$. At each decision point, we keep all the probability distributions for those prices in the base distribution whose values are less than the bid prices, i.e., $s_i \leq \widehat{C_p}(i,t)$. The rest of the distributions are substituted by the following probability representing the likelihood of the out-of-bid event.

$$Pr(C_p(i,t) = \lambda_i) = 1 - \sum_{s_i \leq \widehat{C_p}(i,t)} Pr(C_p(i,t) = s_i) \tag{2.10}$$

Note that it is impossible to generate the precise distribution at each decision point because we do not know the actual realization of the spot price in advance.

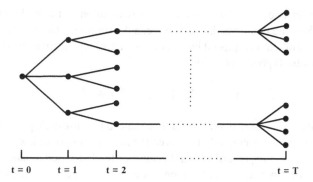

**Fig. 2.5** An example of multistage scenario tree: each leaf vertex represents a scenario, and each non-leaf vertex represents an intermediate state within the planning horizon. A probability is associated with each branch representing the likelihood of state transition

Therefore, the dynamically generated distribution based on the ASP's bid price is an **approximation** to the actual spot price distribution. However, stochastic planning using this approximated distribution outperforms deterministic planning using fixed cost parameters. We will illustrate this point as well as the impact of approximation precision to stochastic planning in the later part of this section.

### 2.4.2.3   Transforming Using Multistage Recourse

We formulate the problem of Stochastic **R**esource **R**ental **P**lanning (SRRP) as a stochastic optimization problem, and build a multistage recourse model to solve this problem. The multistage recourse model allows the application planner to adopt a decision policy that can respond to random events as they unfold. Initially, decisions are made given present resources. As time evolves, possible adjustments (recourse actions) become available to the application planner. As to SRRP, rental planning decisions at various decision points are recourse variables.

The dynamic stochastic spot prices are represented in a multistage scenario tree, $\mathscr{G} = (\mathscr{V}, \mathscr{E})$, presented in Fig. 2.5. A scenario tree has $T + 1$ stages. The first stage represents the current state of the world, and all subsequent stages correspond to the future time slots when new information is available to the application planner. A vertex $v$ in stage $t \in \mathscr{T}$ stands for the state of the system that can be distinguished by information available up to stage $t$. Each vertex $v \in \mathscr{V}$, except the root vertex (indexed as $v = 0$), has a unique parent vertex $\pi(v)$. The probability associated with the state represented by vertex $v$ is $p_v$. Let $\tau(v)$ denote the time stage of vertex $v$ in the tree, we have: $\sum_{\tau(v)=t} p_v = 1$. Each non-leaf vertex $v$ is the root of the subtree: $\mathscr{G}(v) = (\mathscr{V}' \subseteq \mathscr{V}, \mathscr{E}' \subseteq \mathscr{E})$ containing all descendants of vertex $v$. The complete tree is represented by $\mathscr{G} = \mathscr{G}(0)$.

Let the set of leaf vertices of $\mathscr{G}(0)$ be $\mathscr{L}$, and let the set of vertices on the path from the root to vertex $v$ be $\mathscr{P}(v)$. If $v \in \mathscr{L}$, then $\mathscr{P}(v)$ represents a scenario of the problem describing a joint realization of the stochastic parameters over all stages. Otherwise, $\mathscr{P}(v)$ denotes a partial realization of the problem up to the stage $\tau(v)$. With the notations defined above, a decision variable $X_{i,t}$ defined in the deterministic problem is replaced by a set of scenario-dependent decision variables (recourse variables) presented below.

$$X_{i,t} \Rightarrow \{X_{i,v} | \tau(v) = t\}, t \in \mathscr{T} \tag{2.11}$$

The multistage scenario tree is perfectly balanced because each path from root to leaf vertex has the same length $T$. However, the numbers of possible states appeared in each stage are not necessarily equal because of the bid-based dynamic sampling process presented in Sect. 2.4.2.2. Given a scenario tree with a scenario set $S$, the ASP wishes to set a policy that makes different resource rental decisions under different scenarios. For a scenario $S_j \in S$, decisions made at stage $t$ if encountered by scenario $S_j$ is a vector:

$$\{\alpha_{i,v}, \beta_{i,v}, \chi_{i,v}\}, v \in S_j \tag{2.12}$$

The solution must conform to the flow of available information (non-anticipativity). It guarantees that decisions do not rely on information that is not yet available.

### 2.4.2.4  Deterministic Reformulation of SRRP

Having built the multistage recourse model, we derive a deterministic equivalent formulation of SRRP. In the reformulation, the time-dependent decision variables are eliminated. The new formulation introduces a set of new variables that are indexed by the vertices presented in $\mathscr{G}(0)$. Each variable indexed by vertex $v$ is associated with a probability $p_v$. As such, the goal of resource rental planning is to solve MILP with regard to the scenario tree. The complete deterministic equivalent formulation of SRRP is given below:

$$\min \sum_{v \in \mathscr{V}} p_v \cdot (C_f^+(\tau(v)) \cdot \Phi_i \cdot \alpha_{i,v} + (C_s(\tau(v)) +$$

$$C_{io}(\tau(v))) \cdot \beta_{i,v} + C_f^-(\tau(v)) \cdot D(i, \tau(v)) +$$

$$C_p(i, \tau(v)) \cdot \chi_{i,v}) \tag{2.13}$$

s.t.

$$\beta_{i,\pi(v)} + \alpha_{i,v} - \beta_{i,v} = D(i, \tau(v)), \qquad i \in \mathscr{I}, v \in \mathscr{V} \tag{2.14}$$

$$P(i) \cdot \alpha_{i,v} \le Q(i, v), \qquad i \in \mathscr{I}, v \in \mathscr{V} \tag{2.15}$$

$$\alpha_{i,v} \le B \cdot \chi_{i,v}, \qquad i \in \mathscr{I}, v \in \mathscr{V} \tag{2.16}$$

$$\beta_{i,0} = \varepsilon, \qquad\qquad i \in \mathscr{I} \qquad (2.17)$$

$$\alpha_{i,v}, \beta_{i,v} \in \mathbb{R}_+, \qquad\qquad i \in \mathscr{I}, v \in \mathscr{V} \qquad (2.18)$$

$$\chi_{i,v} \in \{0, 1\}, \qquad\qquad i \in \mathscr{I}, v \in \mathscr{V} \qquad (2.19)$$

Since variables at each $t \in \mathscr{T}$ are associated with a number of possible realizations, solving SRRP is equivalent to solving a large-scale MILP. There exist a number of standard techniques to solve this problem, for example, using Benders decomposition [5]. However, due to the huge search space for optimization, they are only suitable for performing short-term resource rental decisions. Fortunately, efficient algorithms are developed that approximate the objective with runtime proportional to the number of nodes on the multistage scenario tree. Readers can refer to Sect. 2.4.2.6 in [32] for more detailed discussions.

### 2.4.2.5 Evaluation of Stochastic Rental Planning Model

In this section, we perform simulations to evaluate the solution to SRRP model. The simulation setting is based on realistic spot pricing history and application-usage scenario presented in Sect. 2.3.3. First, imagine an oracle who knows all the future realizations of spot prices in advance, and takes them as inputs to the DRRP model. We denote the cost generated by this method as the ideal case cost for fine-grained resource rental planning. We then compute the overpay percentages against the ideal case cost for all other approaches. The price distribution is drawn from the same representative data set described in Sect. 2.4.1.2, paragraph 3. The results are plotted in Fig. 2.6. Here, we use the prediction values obtained from the approach described in Sect. 2.4.1 as the bid prices, because they are the best approximation values we can obtain using statistical analysis of past price history. The cost derived by solving SRRP using forecast prices is labeled as "stochastic planning", and the cost of solving its DRRP counterpart and the cost of using on-demand virtual servers are labeled as "predictive planning" and "on-demand-deterministic", respectively. It is not surprising to see that the deterministic planning scheme using on-demand virtual instances yields the most overpay. In addition, stochastic planning is more cost efficient than predictive planning for all three VS types. This is because planning using price distributions is more adaptive to the uncertain availability of spot resources than deterministic planning, and the approximation errors introduced by bidding are "diluted" by fine-grained scenario division at each decision point. When considering the price distribution at every decision point, stochastic planning better hedges against the risk of the unexpected out-of-bid event compared to rental planning based on forecasting values in predictive planning. We also mimic a common bid strategy that ASPs bid a fixed price equal to the expected mean price of the historical data, and compare its cost derived by stochastic and predict planning. The results shown on Fig. 2.6 draw the same conclusion that stochastic planning has better cost advantage.

Next, we investigate the impact of bid price approximation precision to the stochastic planning approach with regard to cost reduction for VS type c1.medium.

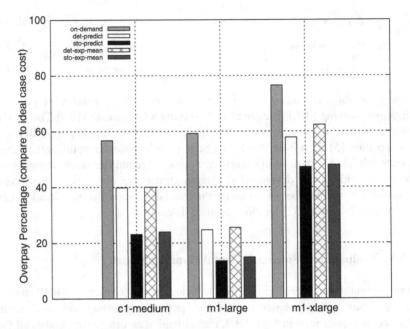

**Fig. 2.6** Comparing predictive and stochastic planning

This evaluation is necessary because according to Sect. 2.4.2.1, the solution quality
of stochastic planning is closely related to the true valuation $\widehat{C_p}(i, t)$, which is
inaccurate in nature with respect to the actual spot price. Taking the cost derived
by actual realization of spot price as the baseline cost, we create artificial bid prices
that are $\pm 2\%$ to $10\%^2$ deviated from the actual price realizations, and measure
the cost deviation from the baseline cost introduced by the approximation errors.
The results converted to percent errors against the baseline cost are plotted in
Fig. 2.7. Clearly, the errors increase as approximation becomes less accurate. We use
the mean squared prediction error (MSPE) to measure the approximation errors.
The MSPE of our best approximation achieved based on the method presented
in Sect. 2.4.1 falls between that of 2 and 4 % deviation of the model. However,
the actual percent error using our approximation is $-12\%$ from the baseline cost.
A possible explanation is that our approximations present a mixture of over- and
under-estimations of the actual price realizations, thus are different from the pattern
of the artificial approximated bid prices we created in the simulation. In conclusion,
if one bids according to the best approximation result in practice, the percentage
error introduced by approximation is generally acceptable.

---

[2]Prices that are more than $\pm 10\%$ from the actual prices are out of the actual price range.

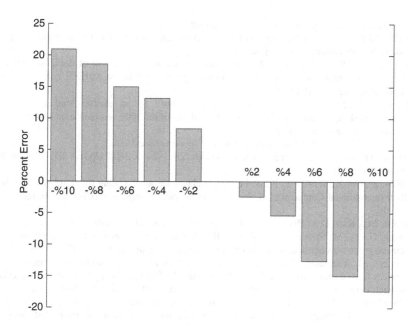

**Fig. 2.7** Impact of approximation precision

# References

1. Agmon Ben-Yehuda, O., Ben-Yehuda, M., Schuster, A., Tsafrir, D.: Deconstructing amazon ec2 spot instance pricing. In: IEEE Third International Conference on Cloud Computing Technology and Science (CloudCom 2011), pp. 304–311 (2011)
2. AIMMS Optimization Software. Available: http://www.aimms.com/
3. Andrzejak, A., Kondo, D., Yi, S.: Decision model for cloud computing under sla constraints. In: Proceedings of the 2010 IEEE International Symposium on Modeling, Analysis and Simulation of Computer and Telecommunication Systems (MASCOTS '10), pp. 257–266 (2010)
4. Berriman, G.B., Deelman, E., Juve, G., Regelson, M., Plavchan, P.: The application of cloud computing to astronomy: A study of cost and performance. CoRR (2010)
5. Birge, J.R.: Decomposition and partitioning methods for multistage stochastic linear programs. Operations Research **33**(5), 989–1007 (1985)
6. Box, G.E.P., Jenkins, G.: Time Series Analysis, Forecasting and Control. Holden-Day, Incorporated (1990)
7. Buyya, R., Ranjan, R., Calheiros, R.: Intercloud: Utility-oriented federation of cloud computing environments for scaling of application services. In: Algorithms and Architectures for Parallel Processing, *Lecture Notes in Computer Science*, vol. 6081, pp. 13–31 (2010)
8. Chaisiri, S., Lee, B.S., Niyato, D.: Optimal virtual machine placement across multiple cloud providers. In: IEEE Asia-Pacific Services Computing Conference (APSCC '09), pp. 103–110 (2009)
9. Chakaravarthy, V.T., Parija, G.R., Roy, S., Sabharwal, Y., Kumar, A.: Minimum cost resource allocation for meeting job requirements. In: 2011 IEEE International Parallel Distributed Processing Symposium (IPDPS '11), pp. 14–23 (2011)
10. Chohan, N., Castillo, C., Spreitzer, M., Steinder, M., Tantawi, A., Krintz, C.: See spot run: using spot instances for mapreduce workflows. In: Proceedings of the 2nd USENIX conference on Hot topics in cloud computing (HotCloud'10), pp. 7–7 (2010)

11. Cloud Exchange. Http://www.cloudexchange.org/
12. IBM ILOG CPLEX optimizer [online].     Available:   http://www-01.ibm.com/software/
    integration/optimization/cplex-optimizer/
13. Deelman, E., Singh, G., Livny, M., Berriman, B., Good, J.: The cost of doing science on
    the cloud: the montage example.   In: Proceedings of the 2008 ACM/IEEE conference on
    Supercomputing (SC '08) (2008)
14. Demberel, A., Chase, J., Babu, S.: Reflective control for an elastic cloud application:
    an automated experiment workbench. In: Proceedings of the 2009 conference on Hot topics in
    cloud computing (HotCloud'09) (2009)
15. EC2 Spot Instance. http://aws.amazon.com/ec2/spot-instances/
16. Market Trends: Platform as a Service, Worldwide, 2012–2016, 2H12 Update. ID: G00239236
    (5 October 2012)
17. Gong, Z., Gu, X., Wilkes, J.: Press: Predictive elastic resource scaling for cloud systems. In:
    2010 International Conference on Network and Service Management (CNSM '10), pp. 9–16
    (2010)
18. Jacob, J.C., Katz, D.S., Berriman, G.B., Good, J.C., Laity, A.C., Deelman, E., Kesselman, C.,
    Singh, G., Su, M., Prince, T.A., Williams, R.: Montage: a grid portal and software toolkit for
    science-grade astronomical image mosaicking. Int. J. Comput. Sci. Eng. **4**, 73–87 (2009)
19. Mattess, M., Vecchiola, C., Buyya, R.: Managing peak loads by leasing cloud infrastructure
    services from a spot market. In: Proceedings of the 2010 IEEE 12th International Conference
    on High Performance Computing and Communications (HPCC '10), pp. 180–188 (2010)
20. Mazzucco, M., Dumas, M.: Achieving performance and availability guarantees with spot
    instances.    In: Proceedings of the 13th International Conference on High Performance
    Computing and Communications (HPCC'11) (2011)
21. Mishra, A.K., Hellerstein, J.L., Cirne, W., Das, C.R.: Towards characterizing cloud backend
    workloads: insights from google compute clusters. SIGMETRICS Perform. Eval. Rev. **37**(4),
    34–41 (2010)
22. Monti, H.M., Butt, A.R., Vazhkudai, S.S.: Catch: A cloud-based adaptive data transfer service
    for hpc. In: 2011 IEEE International Parallel Distributed Processing Symposium (IPDPS '11),
    pp. 1242–1253 (2011)
23. Padala, P., Shin, K.G., Zhu, X., Uysal, M., Wang, Z., Singhal, S., Merchant, A., Salem,
    K.: Adaptive control of virtualized resources in utility computing environments. SIGOPS Oper.
    Syst. Rev. **41**, 289–302 (2007)
24. forecast package for R [online]. Available: http://robjhyndman.com/software/forecast/
25. Song, Y., Zafer, M., Lee, K.W.: Optimal bidding in spot instance market. In: IEEE INFOCOM
    2012, pp. 190–198 (2012)
26. SpotCloud. http://www.spotcloud.com/
27. How to run MapReduce in Amazon EC2 spot market. Available: http://huanliu.wordpress.com/
    2011/06/22/how-to-run-mapreduce-in-amazon-ec2-spot-market/
28. Urgaonkar, B., Chandra, A.: Dynamic provisioning of multi-tier internet applications.   In:
    Proceedings of the Second International Conference on Automatic Computing (ICAC '05),
    pp. 217–228 (2005)
29. Yuan, D., Yang, Y., Liu, X., Chen, J.: A cost-effective strategy for intermediate data storage
    in scientific cloud workflow systems. In: 2010 IEEE International Symposium on Parallel
    Distributed Processing (IPDPS '10), pp. 1–12 (2010)
30. Zhang, J., Kim, J., Yousif, M., Carpenter, R., Figueiredo, R.J.: System-level performance phase
    characterization for on-demand resource provisioning. In: Proceedings of the 2007 IEEE
    International Conference on Cluster Computing (CLUSTER '07), pp. 434–439 (2007)
31. Zhang, Q., Gürses, E., Boutaba, R., Xiao, J.: Dynamic resource allocation for spot markets
    in clouds. In: Proceedings of the 11th USENIX conference on Hot topics in management of
    internet, cloud, and enterprise networks and services (Hot-ICE'11) (2011)
32. Zhao, H.: Exploring Cost-Effective Resource Management Strategies in the Age of Utility
    Computing. Ph.D. thesis, University of Florida, Gainesville, FL, USA (2013)

# Chapter 3
# Efficient and Fair Resource Trading Management

**Abstract** In this chapter, we investigate the resource trading problem in a utility and cloud computing setting where multiple tenants communicate in a Peer-to-Peer (P2P) fashion. Enabling resource trading in cloud unleashes the untapped cloud resources, thus presents a flexible solution for managing resource allocation. However, finding an efficient and fair resource allocation is challenging mainly due to the heterogeneity of resource valuations. Our work first develops a utility-oriented model to support resource negotiation and trading. Based on this model, we adopt a multiagent-based technique that allows a group of autonomous tenants to reach an efficient and fair resource allocation. Further, we add budget limitation to each tenant and propose a directed hypergraph model to facilitate resource trading amongst heterogeneous tenants. We develop a directed hypergraph model to facilitate trading decision making, and design a class of heuristic-based distributed resource trading protocols in favor of different performance metrics.

The rest of the chapter is organized as follows. We first present an overview of the proposed research in Sect. 3.1. We then summarize the related work in Sect. 3.2. In Sect. 3.3, we describe the problem setting and quantify the objectives of the resource trading problem. In Sect. 3.4, we introduce a multiagent-based technique to achieve optimal resource trading efficiency and fairness. Section 3.5 further investigates allocation strategies with limited budget. We propose a novel directed hypergraph model and develop a series of distributed resource trading protocols based on heuristic approaches. Finally, Sect. 3.6 shows simulation results and analyzes their implications.

H. Zhao and X. Li, *Resource Management in Utility and Cloud Computing*,
SpringerBriefs in Computer Science, DOI 10.1007/978-1-4614-8970-2__3,
© The Author(s) 2013

## 3.1 Overview

Nowadays, the utility and cloud computing model is mostly vendor driven, with users having no control over the data or the technology supported by the cloud. Such a vendor-driven model, although convenient to use, brings many issues to light, e.g., failure of monocultures, tradeoff between convenience and control, and concerns about environmental impact [5]. To address these issues, researchers have proposed an alternative model that provides a collaborative resource sharing platform that forms a community-based cloud computing environment [16,18,24]. Different from the centralized vendor model, this community-based cloud leverages under-utilized networked private resources for infrastructure support. Tenants within the same community cloud typically share common security and compliance concerns, and may delegate management to some trusted third-party organization.

Similar to the centralized vendor-driven model, the community-based model offers computation and storage resources as metered services. Therefore, the design goal of the shared cloud resource platform should not only focus on the quality of computing service, but should equally address the economic aspect such that tenants receive cost-effective cloud service provisioning. While managing resource allocation is relatively straightforward in the centralized vendor-driven model (e.g., Amazon®'s on-demand and spot instance pricing), it is particularly challenging due to the heterogeneity in the multitenancy environment. In a community cloud, we are facing a free market where tenants are only incentivized to accept profitable resource exchange. As a result, a well designed multitenancy resource trading protocol is highly desirable to effectively regulate the management of resource allocation.

In this chapter, we study the distributed resource trading problem in a community-based utility and cloud computing environment, and propose a set of multitenancy resource trading protocols to jointly optimize resource allocation efficiency and fairness. Specifically, better efficiency refers to the increased aggregate valuations of all the tenants, and better fairness is interpreted as reduced envy between every pairwise combination of tenants. Our solution follows a market-oriented design principle, and uses a directed hypergraph model to integrate these two seemingly conflicting design objectives into one unified resource trading framework. It directly extends the work of Chevaleyre [10], and further addresses the challenge of budget limited resource trading. With systematic analysis of the resource trading market, a set of heuristic-based distributed resource trading protocols are developed and evaluated.

The comprehensive study presented in this chapter has broad utility in the growing world of "everything-as-a-service". It characterizes the extent to which independent and self-interested tenants interact with each other. Our analysis shows that incentive preserving resource exchanges tend to benefit the system, both from a global view of the overall service efficiency and from a local view of the improved service quality valuation. Moreover, the proposed resource trading approaches are complementary to the vendor-driven cloud computing. For example, consider user

Alice rents a virtual machine from Amazon® with reserved instance pricing. After Alice finishes her job and before the lease expires, Alice might "sublease" this virtual machine to user Bob in order to partially compensate for her resource rental cost.

## 3.2  Related Work

The study described in this chapter presents distributed protocol design to jointly optimize resource trading efficiency and fairness. As the organization of distributed resource evolves towards a more hierarchical architecture [20], distributed algorithms designed for solving combinatorial multi-criteria optimization problems become more attractive. Common optimization techniques include machine learning [26], evolutionary algorithms [13], swarm intelligence [25], and socialeconomy approaches [21, 27]. All these approaches share a common flavor that involves interacting entities evolving towards the optimal solution (by following certain learning or negotiation rules). Our proposed approach falls into the category of socialeconomy approaches. They are built based on the observation that resource management in distributed systems shares common features with commodity allocations driven by market power in the economic study. It is widely adopted to create a computational economy for grid computing [1, 6] and the emerging cloud computing [7, 30]. In an early study, Wolski et al. [32] presented two different market strategies for controlling resource allocation, namely commodities markets and auction. The commodities markets strategy treats disparate resources as interchangeable commodities, while auction requires orchestration from a centralized auctioneer for collecting bids and determining winners. Our proposed resource trading framework is designed for a community cloud environment, and belongs to the commodities market category. In particular, we propose a P2P resource trading market for managing cloud resource allocation. Example research related to this notion includes [12, 31]. In [12], a P2P data replication system was proposed to improve fault-tolerance of digital collections in library. In [31], the authors proposed a multiple currency economy that any peer can issue its own currency. Different from their design, peers directly exchange resources in our distributed resource trading design.

In this chapter, two economic metrics are used to quantify the quality of an allocation: efficiency in terms of overall social welfare, and fairness in terms of envy-freeness. The metric of efficiency is important to characterize the achievable system performance, and was studied in a number of publications [3, 17, 34]. Meanwhile, the metric of fairness highlights individual's utility such that each individual achieves the maximum contentment of its allocated share [14]. Compared to efficiency, the envy-free fairness has generally received far less attentions. A related work targeting grid computing is found in [28]. Using game theory, the authors tackled a multicriteria optimization problem with the aid of axiomatic theory of equity. The authors concluded that for fair and feasible scheduling on global scale computational grid, a strong community control is required. The research conducted in this chapter approaches the multicriteria optimization problem from a different angle, and further investigates how to balance the two metrics amongst budget-aware distributed tenants.

Our proposed protocols utilize a directed hypergraph model. A hypergraph is an extension of the graph concept that one edge (called a hyperedge) can connect an arbitrary set of vertices rather than two. A hypergraph model is flexible and informative to use in algorithm design as it generalizes the graph. For that reason, it becomes attractive to improve algorithm performance in various research domains, e.g., page reputation computation for search engines [2], cellular mobile communication [29] and memory management [19]. For large-scale scientific computing, Çatalyürek and Aykanat [9] proposed a multilevel partitioning approach for mapping repeated sparse matrix-vector computations to multicomputers using hypergraph. Their approach significantly reduces communication overheads while achieving drastically improved mapping results. In their hypergraph model, hyper-edges represent affinity among subsets of the data, and the weights reflect the strength of this affinity. We model the resource trading problem in a similar manner that aims to optimize the aggregate weights of the directed hypergraph model.

## 3.3   A Distributed Resource Trading Framework

This section presents the design overview of a distributed resource trading frame-work for the community cloud. In Sect. 3.3.1, we depict the resource trading system model. In Sect. 3.3.2, we clarify the problem assumptions, define the goals for resource trading, and formulate the problem.

### 3.3.1   System Model

Consider a scenario where a number of highly autonomous tenants connected in a P2P manner, each holding a set of indivisible resources. A resource is an abstraction of hardware bundle or software service, e.g., Virtual Machine (VM), computational time, etc. These resources form a publicly accessible resource pool, and they are completely allocated to all the tenants initially, as described in Fig. 3.1. All tenants form a collaborative community with common purposes and concerns. The underlying P2P communication infrastructure ensures that every tenant is able to talk to every other tenant within the same community (they may not communicate directly, but there is at least one communication path between every pairwise tenants on the topology). For this study, we do not consider dynamic tenants join and leave. We also assume that the distributed system is reliable. Any resource can be assigned to any tenant, incurring certain benefit and cost that may vary depending on the specific resource-tenant assignment. Each tenant can be involved in any number of

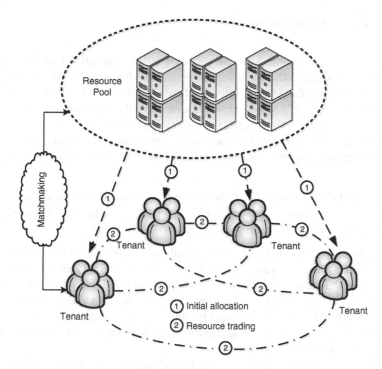

**Fig. 3.1** Multitenancy resource trading: system model

resource trading activities, following the specific tenant negotiation protocol. The distributed resource trading results in a remapping of resources to tenants. We call each instance of such a resource remapping matchmaking. Tenants are incentivized to purchase under-utilized resources from the tenants who currently hold them. As a result, the system evolves towards better resource utilization in the long run.

Formally, let $\mathbb{P} = \{p_1, \ldots, p_n\}$ be the finite set of tenants, and let $\mathbb{R} = \{r_1, \ldots, r_m\}$ be the finite set of indivisible resources. Typically we have $|\mathbb{R}| > |\mathbb{P}|$. This, however, is not necessarily always the case, i.e., some tenants may obtain empty allocation. A matchmaking is defined as a mapping $\mathscr{A}: \mathbb{P} \to 2^{\mathbb{R}}$. More specifically, we have the following definition:

**Definition 3.1 (Matchmaking).** A matchmaking $A = \{A_1, A_2, \ldots, A_n\}$ is a mapping $\mathscr{A}: \mathbb{P} \to 2^{\mathbb{R}}$ satisfying: $A_i \bigcap A_j = \emptyset$, and $\bigcup A_i = A$.

The condition of $\bigcup A_i = A$ ensures that the final matchmaking result is a complete allocation.

### 3.3.2   Problem Statement

For each tenant, we assume a private valuation model, indicating that tenants are mutually blind to each other and evaluate individual allocation independently. The **valuation** of $p_i$ is defined by the valuation function $V_i(\cdot)$, $V_i(\emptyset) = 0$ and $V_i(A_i) \geq V_i(A_i^*)$ for all $A_i \supseteq A_i^*$. Moreover, we assume the valuation function is modular, i.e., $V_i(A_i \cup A_j) = V_i(A_i) + V_i(A_j) - V_i(A_i \cap A_j)$ for all $A_i, A_j \subseteq A$.

Our first goal for distributed resource trading protocol design is to achieve optimal matchmaking efficiency such that the **social welfare**, i.e., $\omega = \sum_{i=1}^{n} V_i(A_i)$, is maximized.

**Definition 3.2 (Efficiency).** Let $\Gamma$ be the set of all possible matchmaking results, an efficient matchmaking is an allocation $A = \{A_1, A_2, \ldots, A_n\}$ that maximizes the social welfare: $\omega_{max} = \max_{A \in \Gamma} \sum_{p_i \in \mathbb{P}} V_i(A_i)$.

The efficiency criterion reflects the overall system performance. For example, suppose there are two resources, one with 2 cores + 1G memory and the other one with 1 core + 2G memory, also assuming user Alice has a CPU-bound job and user Bob has a memory-bound job. Therefore, Alice has higher valuation for the first resource while Bob prefers the second resource. By assigning the first resource to Alice and the second to Bob, the aggregate valuation is maximized, and the system features best job turnaround time.

We define a *resource-bundle* as a collection of one or more resources held by any tenant $p_i$, i.e., a resource-bundle is a non-empty subset of $A_i$. We define a **Deal** as the basic event in the multitenancy resource trading framework. A deal represents the process of resource-bundle transfer from one tenant to another. In order to acquire resources from another tenant, certain amount of compensation is necessary to complete the deal. A **Payment Function** $\varphi_{i,j}$ defines this compensation amount $p_i$ pays to $p_j$. If $\varphi_{i,j}$ is negative, then $p_i$ receives money from $p_j$. Each tenant keeps a record of its payment history. Formally, we define $p_i$'s **Balance** as the summation of its withdrawals and deposits in all deals $p_i$ is involved in: $\theta_i = \sum \varphi_i$. All tenants are utility-driven that seek to make profit at each deal. Formally, suppose after a deal, the allocation of $p_i$ becomes $\tilde{A}_i$, a deal must be a **Rational Deal (RD)** if and only if $V_i(\tilde{A}_i) - V_i(A_i) \geq \varphi_{i,j}$ for all $p_i \in \mathbb{P}$. Note that the requirement of rational deal applies to both tenants involved in the deal, thus is a bilateral constraint. The **Utility** of $p_i$ is given as $U_i(A_i) = V_i(A_i) - \theta_i$.

The second goal of our protocol design is to promote fairness within the system. By associating the valuation and payment function, fairness denotes to envy-free [4] amongst all tenants, indicating that no tenant would get better off by swapping its allocation with another peer though a rational deal. Specifically, the definition of a fair allocation is given as follows.

**Definition 3.3 (Fairness).** Let $\Gamma$ be the set of all possible matchmakings, a matchmaking result is characterized as fair iff there exists $A = \{A_1, A_2, \ldots, A_n\} \in \Gamma$ such that: (a) $\forall p_i, p_j \in \mathbb{P}$, $p_i$ and $p_j$ has direct connection; and (b) $V_i(A_i) - \theta_i \geq V_i(A_j) - \theta_j$.

The fairness criterion is in line with the envy-free definition given out in [10] that takes transferable utility into account. The authors proved that a Efficient and Envy-Free (**EEF**) state always exists. Here, we further extend their result by adding topology constraint to the fairness definition. Our definition limits envy-free states to neighboring tenants. This is justifiable as the underlying communication topology might not be a fully connected network. In addition, a common practice in distributed systems is to employ a budget transfer mechanism to enforce incentives for community control [22]. For example, in P2P and social networks, some form of digital cash, or numerical reputations representing trust relationships may be used for rewarding and punishing certain actions. We formally define budget constraint as follows.

**Definition 3.4 (Budget).** Budget $b_i^t$ expresses maximum amount $p_i$ is able to offer after $t$ deals. Let $b_i^0$ be the initial budget initially, we have:

$$b_i^t = b_i^0 - \theta_i^t$$

Given any initial allocation, the goal of this study is to investigate to what extent efficiency and fairness can be achieved in the multitenancy resource trading framework described above, and to design resource trading protocols to guide tenant interactions evolving towards system-wide efficiency and fairness. We analyze situations with and without the budget limitation. From now on, we label the scenario with budget constraint as **budget-aware**, and refer to the later scenario as **budget-unaware**.

## 3.4 Budget-Unaware Resource Trading Protocol

In this section, we develop a resource trading protocol without the presence of budget constraint. Our protocol design is based on the multiagent-based resource allocation optimization framework presented in [10].

### 3.4.1 Preliminaries

By following certain payment rules, we will show that the resource trading protocol is capable of reaching topology-wide efficiency as well as envy-free fairness upon convergence. A topology-wide efficient allocation is an allocation such that for every tenant, the allocation for the sub-topology consisting of that tenant and its direct neighbors is efficient, i.e., the matchmaking achieves maximum social welfare on the sub-topology. We introduce topology-wide efficiency because for a partially connected communication topology, a globally efficient matchmaking is not guaranteed unless the order of resource trading is carefully planned. An example is given out in [33], Sect. 3.4.1.

In the resource trading framework, each tenant completes transactions with neighbors using only rational deals (RD), and obtains or loses resource bundle accordingly. An RD indicates that the transaction is beneficial for pushing resources to tenants who value them more. In fact, **ANY** sequence of RD executions will achieve efficiency with regard to the underlying communication topology. This is due to the following observations: (1) RD increases social welfare according to its definition; and (2) if no more RD is possible, then the matchmaking must reach the maximum possible social welfare. Given modular valuation function, we have the following proposition.

**Proposition 1 (Convergence to Efficiency [15]).** *Any sequence of RD involving any number of resource exchanges will eventually yield to topology-wide efficiency.*

The reasoning behind Proposition 1 is fairly simple. Each RD results in remapping of resources to tenants with higher interests. When no RD is possible with respect to the communication topology, the system converges to a topology-wide efficient state. Another implication is that the final state is independent of the execution order of RDs. Now suppose after an execution of an RD, the current allocation becomes $\tilde{A}$. Since the deal is bilaterally beneficial to both tenants involved in the deal, we calculate the payment range with the following equations.

$$V_i(\tilde{A}_i) - V_i(A_i) \geq \varphi_{i,j}$$
$$V_j(\tilde{A}_j) - V_j(A_j) \geq -\varphi_{i,j} \tag{3.1}$$

By solving this equation, the result of the payment function $\varphi_{i,j}$ falls into the range of $[V_j(A_j) - V_j(\tilde{A}_j), V_i(\tilde{A}_i) - V_i(A_i)]$, i.e., the rational payment range.

### 3.4.2   A Multiagent Based Optimization Approach for Resource Trading

This section introduces the theoretical foundation of our multitenancy resource trading protocol design. It is mainly based on the theoretical framework developed by Chevaleyre et al. [10, 11] for multiagent systems. One central conclusion is that resource allocation efficiency and fairness can be simultaneously achieved in a multiagent negotiation framework. In order to achieve this state, a proper payment function was selected to deal with the increased social surplus $\omega(\tilde{A}) - \omega(A)$ after each deal. In particular, a payment function called **Globally Uniform Payment Function (GUPF)** was proposed. Suppose $A$ and $\tilde{A}$ are allocations before and after an RD execution, respectively, the GUPF is defined as follows.

$$\textbf{GUPF: } \varphi_i = [V_i(\tilde{A}_i) - V_i(A_i)] - \frac{[\omega(\tilde{A}) - \omega(A)]}{n} \tag{3.2}$$

Equation (3.2) is labeled as globally uniform because this payment is imposed on all tenants. For tenants who do not involved in the deal, $V_i(\tilde{A}_i) - V_i(A_i)$ equals to zero, so each of them receives an equal share of the social surplus created by the trading activity. Note that GUPF is within the bound of rational payment (3.1). In addition to GUPF, a one-off payment amount at initial is introduced. The initial payment amount, called **initial equitability payment**, is defined as: $\varphi^0 = V_i(A_i^0) - \frac{\omega(A^0)}{n}$. The main purpose for this payment function is to "level the playing field". The next two theorems show that imposing initial equitability payment and GUPF for resource trading leads to efficient and fair matchmaking. The following theorem shows that individual utility is invariant after every RD.

**Theorem 3.1.** *If each tenant pays initial equitability payments at start and pays GUPF after each RD executes, then all tenants share the same utility: $U_i(A_i) = \frac{\omega(A)}{n}$ after each RD.*

With this invariant, we prove the following theorem. Note that our version is slightly different from that presented in [10], as we target at topology-wide efficiency and use a more strict assumption of modular domain.

**Theorem 3.2 (Convergence to Efficiency and Fairness [10]).** *When all valuations are modular and budget limitation is not a concern, paying initial equitability payment at start and GUPF after each RD for every $p_i \in \mathbb{P}$ will converge to a matchmaking state that achieves both topology-wide efficiency and envy-free fairness.*

More details about these two theorems and the implementation of the protocol are described in the Sect. 3.4 of [33].

## 3.5 Budget-Aware Resource Trading Protocol

### 3.5.1 Modeling Resource Trading Using a Directed Hypergraph

When budget constraint is imposed, the convergence to the optimal matchmaking state might not exist. In this section, we develop a directed hypergraph model for community-based cloud resource trading. A hypergraph is a generalization of the 2D graph that an edge can connect a set of vertices. If the hypergraph is directional, an edge (a.k.a. a hyperarc) connects a hypernode (head) with a set of hypernodes (tail set). The motivation behind the directed hypergraph model lies in its implication for one-to-many relationship. A 2D graph merely models connectivity among tenants, but cannot represent task allocation and envy relationship among them. A directed hypergraph is more informative, succinctly capturing the scenario that a resource is held by some tenant, but inspires more interest from some other tenants each holding a set of resources.

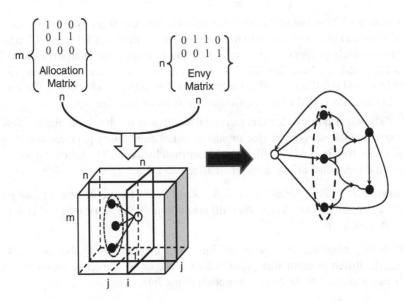

**Fig. 3.2** A directed hypergraph model. The proposed directed hypergraph model derives from an $m \times n \times n$ hyperspace. A hypernode is a point mapping allocation and envy relationship on the hyperspace. A hyperarc connects a hypernode $v \in V$ with a set of other hypernodes belonging to a common tenant and has envy relationship with the v's host tenant. An example directed hypergraph is shown on the *right side*

We propose two matrices to build up a hyperspace. The first matrix is an Allocation Matrix (AM). It is an $m \times n$ matrix that takes binary values, representing current resource matchmaking state for all tenants. Each entry $\alpha_{i,j}$ in AM is defined as follows.

$$\alpha_{i,j} = \begin{cases} 1 & \text{tenant } j \text{ holds resource } i \\ 0 & \text{otherwise} \end{cases}$$

The second matrix is an Envy Matrix (EM) representing current matchmaking unfairness (or envy relationship). Suppose we have two tenants, Alice and Bob. Bob is said to envy Alice when Bob has higher valuation for some resource currently allocated to Alice. Again, we use binary values to represent the envy relationships. Formally, An Envy Matrix is a $n \times n$ matrix defined as follows.

$$\varepsilon_{i,j} = \begin{cases} 1 & p_i \text{ is envies } p_j \\ 0 & \text{otherwise} \end{cases}$$

Combining the allocation matrix and the envy matrix, we are ready to unifying allocation and envy relationships into one directed hypergraph model. We first create a three-dimensional space, $m \times n \times n$, as shown in the left side of Fig. 3.2. A directed hypergraph $H = (V, E)$ is composed of a finite non-empty set $V$ of hypernodes and a finite non-empty set $E$ of hyperarcs. Using the coordinates of the hyperspace, we define the hypernode as follows. A hypernode $v$ (3.3) is a three-tuple $(x, y, z)$, where $x \in \mathbb{R}$ represents the resource, $y \in \mathbb{R}$ represents the tenant currently holding $x$, and $z$ is some tenant has envy relationship with $y$, i.e., $z \in \mathbb{P}, \varepsilon_{y,z} = 1$. A hyperarc $e$ (3.4) is a pair $< T, h >$, where $T \subseteq V$ is the tail of $e$ and $h \in V \setminus T$ is its head. The tail set $T$ includes those hypernodes whose host tenants involved in envy relationships with the host of the head.

**Hypernode**: A hypernode is a three-tuple:

$$v = (x, y, z) \in V$$

$$\text{s.t. } x \in \mathbb{R}$$

$$y \in \mathbb{P}, \text{and } \alpha_{x,y} = 1$$

$$z \in \mathbb{P}, \text{and } \varepsilon_{y,z} = 1 \tag{3.3}$$

**Hyperarc**: A hyperarc $e \in E$ is an ordered pair $< T, h >$ iff:

$$e =< T, h >\in E$$

$$\text{s.t. } h = (x_1, y_1, z_1) \in V$$

$$v = (x_2, y_2, z_2) \in T \subseteq V$$

$$y_2 = z_1 \tag{3.4}$$

Each tenant can establish a local view of the directed hypergraph. The hyperarcs imply potential transactions to be negotiated. In a distributed environment, when one transaction is accomplished using an RD, resource allocation changes which might affect other resource trading activities. Building a directed hypergraph is thus helpful to evaluate the quality of trading selections. For example, there are many applications of the optimal structures in the proposed directed hypergraph model, such as optimal spanning hypertree and optimal edge cover. Readers can find more information in Sect. 3.5.2 of [33].

### 3.5.2 Protocol Design

When proposing for resource trade, a tenant rationally calculates its payment amount. When the budget limitation $b_i$ is imposed on $p_i \in \mathbb{P}$, the rational payment amount $\varphi_{i,j}$ for trade proposal is in the range of:

$$\varphi_{i,j} \in [V_j(A_j) - V_j(\tilde{A}_j), \min\{V_i(\tilde{A}_i) - V_i(A_i), b_i\}]. \tag{3.5}$$

---

**Protocol 1:** (a) V-BaMRT (b) E-BaMRT (c) P-BaMRT

---

**begin**

  **for** $p_i \in \mathbb{P}$ **do**

    Establishes local view with neighboring peers;

    // --Trade Proposal--

    **while** $p_i$ *has at least one envious neighbor* **do**

      a) Sorts potential transactions based on envy degree;

      b) Selects $p_j$ with the highest envy degree drop;

      c) Selects payment within the range defined by Equation 3.5;

      **if** $p_j$ *accepts offer* **then**

        • Make payments;

        • Removes $p_j$ from its envy list

    // --Offer Selection--

    **while** *conflicting offers arrival* **do**

      Selects offer with;

$$\begin{cases} \text{(a) highest social welfare gain, or} \\ \text{(b) largest envy degree decrease, or} \\ \text{(c) highest transaction profits} \end{cases}$$

    Accepts offer;

    Receives payments and updates local view;

---

According to analysis in Sect. 3.4.2, resource allocation in the community cloud evolves towards efficient and fair state when tenants pay initial equitability $\varphi^0$ and GUPF in BuMRT. However, when budget limitation presents, tenants do not always abide by these routine payments. Therefore, we are interested in investigating the transition of resource allocation states, when tenants pay different amounts as long as the amounts fall in the range of (3.5). In this section, we propose a series of heuristic-based BaMRTs. These protocols confine the trading activities of each tenant to neighboring peers, allowing them to conduct local negotiations. However, they are different with each other in terms of trading selection criterion. The complete description of the proposed BaMRT protocols are illustrated in Protocol 1.

Tenants delegate trading controls to trading agents who perform two basic operations periodically: proposing trade and selecting offer. When proposing a trade, the agent simply selects the neighboring peer who he envies most as the trading partner. In order to quantify the matchmaking unfairness between pairwise trading partners, we use the following equation to define the envy degree on a particular hyperarc.

$$\sigma_{i,j} = \max\{U_i(\tilde{A}_i, \tilde{\theta}_i) - U_i(A_i, \theta_i), 0\}$$

The trading agent may select any payment amount within the rational range. A tenant can set up a predefined payment policy for the trading agent. For example, a conservative policy results in resource acquisition with low cost, while an aggressive policy helps funding peer tenants to conduct further trades, and might benefit more in return. We will evaluate different payment policies in the performance evaluation section. When multiple offers arrive, each trading agent needs to carefully evaluate trading decisions with a local view of the directed hypergraph model. This is especially important when offers conflict with each other since the resource can only be granted to one neighbor. In our design, each trading agent employs a hill climbing technique to negotiate resource trading with neighboring peers. The hill climbing algorithm is fast and effective in finding a local optimal matchmaking. The local optimal offer selection decision must be rational, as the payment amounts conforming to RD increase the overall social welfare (Proposition 1). In other words, if a trade occurs, the allocation efficiency is reinforced, and the corresponding envy relationship between the trading parties is eliminated.

We propose three versions of BaMRT in favor of different trading selection criterion. Each of them follows different paths to reach the local minimum. The first version labeled as **Valuation oriented BaMRT (V-BaMRT)**, let trading agents select trades with the highest social welfare gain. In the second version, each agent selects the neighboring peer who he envies most as the trading partner. We label this version of BaMRT as **Envy oriented BaMRT (E-BaMRT)**. Finally, we propose **Profit oriented BaMRT (P-BaMRT)**, in which agents select offers that will bring in the highest transaction profits (defined as the difference of payment and gained valuation). These protocols work similarly to BuMRT except that they do not require message broadcasting to redistribute social wealth within the community.

## 3.6   Performance Evaluation

In this section we investigate the performance of the proposed protocols through three different sets of simulations. First, we implement BuMRT and validate its achievable efficiency and fairness. In the second set of simulations, three versions of BaMRT presented in Protocol 1 are compared in various norms. Finally, we evaluate the performance impact of different payment selection policy and initial budget settings for BaMRT.

### 3.6.1   Simulation Settings

We instantiate the matchmaking framework to a generalized distributed computing environment, and implement the resource trading protocols using SimGrid [8]. The core scheduling and communication functions are implemented using the application-level simulation interfaces provided by the MSG module of SimGrid.

A community cloud platform with 20 computational nodes (tenants) is simulated. We also creates 800 synthetic task units (resources). To create a heterogeneous platform, we assign different computational and networking settings to the computational nodes. As such the same task unit presents different values to different nodes. In SimGrid, this information is encapsulated in separate XML files. Node $i$'s satisfaction of its current allocation is quantified by a concave valuation function $V_i(\cdot)$, where $V_i(x)$ defines the utility of node $i$ obtaining $x$ tasks [23]. The concavity assumption indicates that the marginal valuation diminishes when the allocation increases. Specifically, we use the following concave function to represent valuation,

$$V_i(x) = c \cdot x^r,$$

where the constant coefficient $c$ is set to 10.0, and $r$ is randomly generated in the range of $(0.2, 0.6)$.

We primarily use four metrics to evaluate the performance of the proposed protocols. First, we use *social welfare* to quantify the allocation efficiency. Second, in order to validate fairness, the *total envy degree* amongst all nodes is recorded after each transaction. In addition, two nodes that envy each other form an envious pair. The total number of envious pairs is also counted throughout the negotiation process. Finally, we measure *system profit* as an indication of system's side utility. For each transaction, the profit earned is the difference of buyer's valuation and the associated payment amount. The system profit is thus defined as the cumulative profit earned in all transactions.

### 3.6.2   Evaluation of BuMRT

In the first set of simulations, nodes negotiate with each other using BuMRT until convergence is reached. The results are plotted in Fig. 3.3. At the start of each simulation, 800 task units are randomly mapped to 20 nodes. We generate three topology profiles representing different network configurations. The first topology profile (labeled as "fully connected") describes a fully connected mesh network, and the rest profiles describe two relatively sparse network topologies. The fully connected topology has a total node degree of $20 \times 19 = 380$. The node degrees of the other two profiles are normalized relative to the fully connected profile. We use these normalized values, 0.45 and 0.72, to represent the connectivity of both profiles. In order to validate efficiency, we also implement a self-adaptive auction algorithm [34] that achieves maximum social welfare when tasks are allocated. This result, labeled as "optimal" in Fig. 3.3a, defines the global optimal social welfare. From Fig. 3.3a, we observe that in all topology profiles, the overall social welfare increases all the time and converges after around 24 transactions. In addition, for the fully connected network, the final allocation achieves the maximum social welfare when converges. Figure 3.3b, c show that all simulations converge to fair state where all envy relations are eliminated. Note that after each

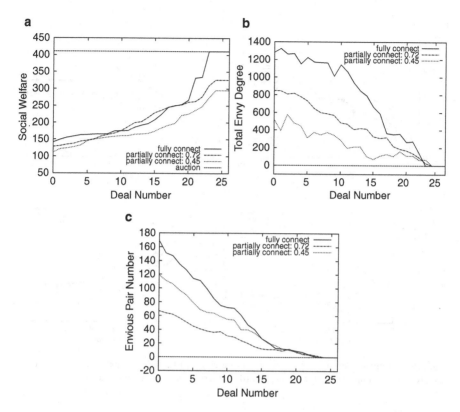

**Fig. 3.3** Performance evaluation for budget-unaware case. (**a**) Measurement of efficiency; (**b**) measurement of fairness: envy degree; (**c**) measurement of fairness: envious pair

transaction, both envy degree and envious pair number do not necessarily decrease. This can be explained as follows: although the overall unfairness will be eliminated eventually, each single transaction only eliminates envy between the two trading partners, but may create envy relationship between other pairs. Another interesting observation for Fig. 3.3 is that the initial matchmaking unfairness is closely related to the network connection degree. This is because envy relation is more likely to present if more nodes are connected. Moreover, more connected network also means more opportunities for tasks to be assigned to nodes who value them more. Therefore, the achievable local efficiency is more likely to increase as the network becomes more connected.

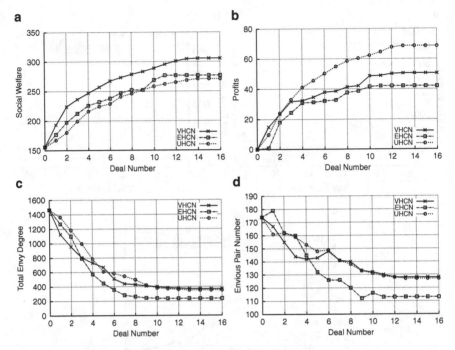

**Fig. 3.4** Performance evaluation for budget-aware case. (**a**) Efficiency improvement; (**b**) profits gain; (**c**) fairness improvement: envy degree; (**d**) fairness improvement: envious pair

### 3.6.3   Evaluation of BaMRT

Next, we add budget limitation to each node and compare the performance of different versions of BaMRT presented in Sect. 3.5. The node and the task unit number are set to be 20 and 800 respectively. Based on the analysis of the average transaction payment range, we assign each node an initial budget of 100. When a transaction is completed, the node who makes payment will deduct the corresponding amount from its balance. Conversely, its trade partner will add the same amount to its balance. For fair comparison, all simulations are conducted using the same setting (valuation functions and initial allocation). All simulations use a same fully-connected network. The comparison results are exhibited in Fig. 3.4. From these results, we draw the conclusion that the performance of each protocol is primarily influenced by the offer selection strategy. In V-BaMRT, the offer brings the most social welfare growth is selected. Therefore in Fig. 3.4a we observe that V-BaMRT leads to the highest local efficiency when converged. Similarly, Fig. 3.4c, d show that E-BuMRT performs better in promoting fairness. And not surprisingly, the overall profits gain is in favor of U-BuMRT, as shown in Fig. 3.4b.

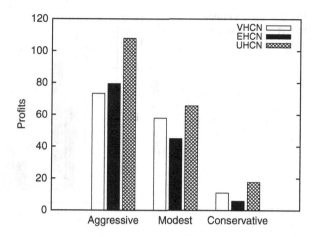

**Fig. 3.5** Impact of different payment selection strategies for budget-aware case

## 3.6.4   Sensitivity Analysis

In this section, we investigate the impact of different payment selection strategies and initial budget settings. The configuration parameters are kept the same as in Sect. 3.6.3.

As analyzed in Sect. 3.5.2, each tenant can set up arbitrary payment policy for the trading agent. A conservative policy results in resource acquisition with low cost, while an aggressive policy helps funding other tenants to conduct more trading activities. Which policy gives better result depends on the offer selection strategies and initial budget distribution. We modify the simulation code to let each node select payment amount within the allowed range deterministically. Specifically, let the payment selection range be $(low, high)$, we devise three deterministic payment selection strategies for evaluation:

- Aggressive: $payment = low + 0.75 \times (high - low)$
- Modest: $payment = low + 0.5 \times (high - low)$
- Conservative: $payment = low + 0.25 \times (high - low)$

We compare the aggregate profits of the system in Fig. 3.5. Each value is the average result of 20 simulation runs. The result suggests that more aggressive bidding behavior will result in higher system profits at convergence. This can be explained that if all nodes offer higher at each deal, more nodes will get funded that lead to more transactions. As a result, the micro-economy of the small computing community is boosted.

Finally, we alter the initial budget assignment and measure its impact to the system envy degree. Taking initial budget of 100 to be the base case (marked as "1X"), the startup fund for each node is altered from 0.5 to 2 times of 100. Again we average the result of 20 simulation runs. The comparison is visualized in Fig. 3.6. We observe that for the case of abundant initial fund assignment, the convergence value is close to that achieved by BuMRT. When the initial budget reaches 200,

**Fig. 3.6** Impact of initial
budget assignment for
budget-aware case

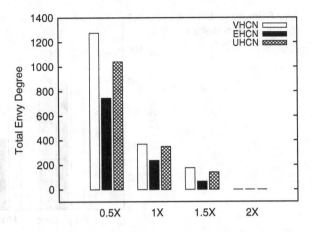

all protocols converge to an envy degree of 0 as if there are no budget constraint.
On the contrary, for a poorly funded computing community, the trading activities are
more likely to freeze due to lack of budget, resulting in potential longer convergence
time and higher degree of unfairness.

# References

1. Abramson, D., Buyya, R., Giddy, J.: A computational economy for grid computing and its
   implementation in the nimrod-g resource broker. Future Gener. Comput. Syst. **18**, 1061–1074
   (2002)
2. Berlt, K., Moura, E.S.D., Carvalho, A.L.D.C., Cristo, M., Ziviani, N., Couto, T.: A hypergraph
   model for computing page reputation on web collections. In: Brazilian Symposium on
   Databases, pp. 35–49 (2007)
3. Boche, H., Naik, S., Alpcan, T.: Characterization of non-manipulable and pareto optimal
   resource allocation strategies for interference coupled wireless systems. In: Proceedings of the
   29th conference on Information communications (INFOCOM 2010), pp. 2417–2425 (2010)
4. Brams, S.J., Taylor, A.D.: Fair division: from cake-cutting to dispute resolution. Cambridge
   University Press (1996)
5. Briscoe, G., Marinos, A.: Digital ecosystems in the clouds: Towards community cloud
   computing. In: 3rd IEEE International Conference on Digital Ecosystems and Technologies
   (DEST'09)., pp. 103–108 (2009)
6. Buyya, R., Abramson, D., Venugopal, S.: The grid economy. Proceedings of the IEEE **93**(3),
   698–714 (2005)
7. Buyya, R., Pandey, S., Vecchiola, C.: Cloudbus toolkit for market-oriented cloud computing.
   In: Proceedings of the 1st International Conference on Cloud Computing (CloudCom'09),
   pp. 24–44 (2009)
8. Casanova, H., Legrand, A., Quinson, M.: SimGrid: a Generic Framework for Large-Scale
   Distributed Experiments. In: 10th IEEE International Conference on Computer Modeling and
   Simulation (UKSIM 2008) (2008)
9. Çatalyürek, U., Aykanat, C.: Hypergraph-partitioning-based decomposition for parallel sparse-
   matrix vector multiplication. IEEE Trans. Parallel Distrib. Syst. **10**(7), 673–693 (1999)

10. Chevaleyre, Y., Endriss, U., Estivie, S., Maudet, N.: Reaching envy-free states in distributed negotiation settings. In: Proceedings of the Twentieth International Joint Conference on Artificial Intelligence (IJCAI 2007), pp. 1239–1244 (2007)
11. Chevaleyre, Y., Endriss, U., Maudet, N.: Allocating goods on a graph to eliminate envy. In: Proceedings of the 22nd AAAI Conference on Artificial Intelligence (AAAI 2007), pp. 700–705 (2007)
12. Cooper, B., Garcia-Molina, H.: Peer-to-peer resource trading in a reliable distributed system. In: the First International Workshop on Peer-to-Peer Systems (IPTPS 2001), pp. 319–327 (2002)
13. di Costanzo, A., de Assuncao, M.D., Buyya, R.: Harnessing cloud technologies for a virtualized distributed computing infrastructure. IEEE Internet Computing 13, 24–33 (2009)
14. Edmonds, J., Pruhs, K.: Balanced allocations of cake. In: Proceedings of the 47th Annual IEEE Symposium on Foundations of Computer Science (FOCS 2006), pp. 623–634 (2006)
15. Estivie, S., Chevaleyre, Y., Endriss, U., Maudet, N.: How equitable is rational negotiation? In: Proceedings of the fifth international joint conference on Autonomous agents and multiagent systems (AAMAS 2006), pp. 866–873 (2006)
16. Eucalyptus Community Cloud. http://www.eucalyptus.com/eucalyptus-cloud/community-cloud
17. Gradwell, P.: Distributed combinatorial resource scheduling. In: 1st International Workshop on Smart Grid Technologies (2005)
18. Federal    Community    Cloud.    http://www-304.ibm.com/industries/publicsector/us/en/contenttemplate1/!!/xmlid=207581
19. Krishnamoorthy, S., Çatalyürek, U., Nieplocha, J., Rountev, A., Sadayappan, P.: Hypergraph partitioning for automatic memory hierarchy management. In: Proceedings of the 2006 ACM/IEEE conference on Supercomputing (SC 2006) (2006)
20. Kwok, Y.K., Hwang, K., Song, S.: Selfish grids: Game-theoretic modeling and nas/psa benchmark evaluation. IEEE Trans. Parallel Distrib. Syst. 18, 621–636 (2007)
21. Lai, K., Rasmusson, L., Adar, E., Zhang, L., Huberman, B.A.: Tycoon: An implementation of a distributed, market-based resource allocation system. Multiagent Grid Syst. 1, 169–182 (2005)
22. Laoutaris, N., Poplawski, L.J., Rajaraman, R., Sundaram, R., Teng, S.H.: Bounded budget connection (bbc) games or how to make friends and influence people, on a budget. In: Proceedings of the twenty-seventh ACM symposium on Principles of distributed computing (PODC 2008), pp. 165–174 (2008)
23. Ma, R.T., Chiu, D.M., Lui, J.C., Misra, V., Rubenstein, D.: On resource management for cloud users: A generalized kelly mechanism approach. Tech. rep., Electrical Engineering (2010)
24. Marinos, A., Briscoe, G.: Community cloud computing. CoRR **abs/0907.2485** (2009)
25. Pandey, S., Wu, L., Guru, S.M., Buyya, R.: A particle swarm optimization-based heuristic for scheduling workflow applications in cloud computing environments. In: Proceedings of the 24th IEEE International Conference on Advanced Information Networking and Applications (AINA 2010), pp. 400–407 (2010)
26. Perez, J., Germain-Renaud, C., Kégl, B., Loomis, C.: Multi-objective reinforcement learning for responsive grids. Journal of Grid Computing 8, 473–492 (2010)
27. Regev, O., Nisan, N.: The POPCORN market. online markets for computational resources. Decision Support Systems 28(1–2), 177–189 (2000)
28. Rzadca, K., Trystram, D., Wierzbicki, A.: Fair game-theoretic resource management in dedicated grids. In: Proceedings of the Seventh IEEE International Symposium on Cluster Computing and the Grid (CCGrid 2007), pp. 343–350 (2007)
29. Sarkar, S., Sivarajan, K.: Hypergraph models for cellular mobile communication systems. IEEE Transactions on Vehicular Technology 47(2), 460–471 (1998)
30. Song, B., Hassan, M., Huh, E.N.: A novel cloud market infrastructure for trading service. In: Proceedings of the 2009 International Conference on Computational Science and Its Applications (ICCSA 2009), pp. 44–50 (2009)
31. Turner, D.A., Ross, K.W.: A Lightweight Currency Paradigm for the P2P Resource Market. In: Seventh International Conference on Electronic Commerce Research (2004)

32. Wolski, R., Plank, J.S., Brevik, J., Bryan, T.: G-commerce: Market formulations controlling resource allocation on the computational grid. In: Proceedings of the 15th International Parallel & Distributed Processing Symposium (IPDPS 2001), pp. 46–53 (2001)
33. Zhao, H.: Exploring Cost-Effective Resource Management Strategies in the Age of Utility Computing. Ph.D. thesis, University of Florida, Gainesville, FL, USA (2013)
34. Zhao, H., Li, X.: Efficient grid task-bundle allocation using bargaining based self-adaptive auction. In: Proceedings of the 9th IEEE/ACM International Symposium on Cluster Computing and the Grid (CCGrid 2009), vol. 0, pp. 4–11 (2009)

# Chapter 4
# Flexible Resource Sharing Management

**Abstract** This chapter presents CloudBay, an online resource trading and leasing platform for multi-party resource sharing. It is a proof-of-concept design bridging the gap between resource providers and resource customers. With the help of Cloud-Bay, the untapped computing power privately owned by multiple organizations is unleashed. The design and implementation of the CloudBay project presents the most challenge to our exploration of cost-effective resource management strategy design. Following a market-oriented design principle, CloudBay provides an abstraction of a shared virtual resource space across multiple administration domains, and features enhanced functionalities for scalable and automatic resource management and efficient service provisioning. CloudBay distinguishes itself from existing research and contributes in mainly two aspects. First, it leverages scalable network virtualization and self-configurable virtual appliances to facilitate resource federation and parallel application deployment. Second, CloudBay adopts an eBay-style transaction model that supports differentiated services with different levels of job priorities. For cost-sensitive users, CloudBay implements an efficient match-making algorithm based on the auction theory and enables opportunistic resource access through preemptive service scheduling. The proposed CloudBay platform stands between HPC service sellers and buyers, and offers a comprehensive solution for resource advertising and stitching, transaction management, and application-to-infrastructure mapping. In this chapter, we present the design details of CloudBay, and briefly discuss lessons learnt and challenges encountered in the implementation process.

The rest of the chapter is organized as follows. We will first present an overview for the CloudBay project in Sect. 4.1. Next, we survey the related system design and implementation in Sect. 4.2. In Sect. 4.3, we describe the design of CloudBay and the implementation of network virtualization tools facilitating resource sharing. In Sect. 4.4, we explain the details of the job scheduling algorithms in CloudBay.

H. Zhao and X. Li, *Resource Management in Utility and Cloud Computing*,
SpringerBriefs in Computer Science, DOI 10.1007/978-1-4614-8970-2_4,
© The Author(s) 2013

Finally, the evaluation results of our prototype CloudBay implementation are presented in Sect. 4.5. The original text of this chapter was published in [37], and we refer the readers to [36] for more detailed discussion of the subject.

## 4.1  Overview

Utility and cloud computing reshape the way IT services are delivered with its ability to elastically grow and shrink the resource provisioning capacity on demand. This computing paradigm shift launches a new chapter for e-science and e-engineering applications that offers High Performance Computing (HPC) at scale. For example, the recent published top500 list includes Amazon®'s EC2 virtual cluster composed of over one thousand *cc2.8xlarge* instances [16]. In order to realize HPC-as-a-service with the full potential of utility and cloud computing, it is best to take advantage of resources in an open marketplace across multiple clouds [18]. However, two major challenges still remain to be addressed. First, although end users are liberated from the arduous task of resource configuration, this burden is transferred to computational resource providers. Existing work either limits service to local area connectivity [1], or requires nontrivial resource and networking setup among all resource contributors [8]. Second, there lacks a flexible application-to-infrastructure mapping mechanism that accommodates differentiated service requirements, and at the same time, maintains high efficiency for resource allocation across multiple clouds. Finally, it is critical to implement a fair pricing scheme in a multi-party cloud computing environment for both resource sellers and customers.

To overcome these hurdles, we propose a proof-of-concept design, CloudBay, as a full-fledged solution for computational resource sharing and trading in an open cloud environment. CloudBay addresses the first challenge by incorporating decentralized self-configurable networking and self-packaging cloud toolsets. This design breaks the barrier of proprietary clouds and reduces efforts for resource joining, maintenance and query. It also helps cloud resource customers to deploy and maintain their applications using the shared cloud infrastructure. To address the second challenge, CloudBay implements an eBay-style transaction model. Specifically, user requests are classified as quality-sensitive and cost-sensitive depending on the offers the users are willing to make. The service scheduler in CloudBay assigns higher priorities to quality-sensitive service requests, and allows opportunistic provisioning of under-utilized resources through preemptive application execution. The competition among cost-sensitive service requests are resolved by an efficient auction mechanism that guarantees resource access for those users who value them the most. Service scheduling in CloudBay also supports distributed sites bid for jobs for optimal system-wide performance. Our proposed market-driven solution differs from Amazon®'s on-demand and spot IaaS in the following two aspects: (1) CloudBay allows user requests to be partially fulfilled, whereas EC2 spot market only supports all-or-none resource acquisition. This feature is useful as HPC users often have fuzzy resource demand [4]. (2) Resource auction in CloudBay is based on a novel Ausubel auction model [3] that encourages truthful bidding (i.e., bidders bid based on their true valuation), and achieves Vickrey efficiency compared with Amazon®'s spot market auction. The development of CloudBay is still in progress and more features will be added in future. We believe that the exploratory investigation presented in this study can open up significant perspectives of merging HPC and cloud computing in the long run.

In summary, we demonstrate that the following features render CloudBay a favorable design for HPC-as-a-service in an open cloud environment:

- **Scalable resource federation:** Leveraging P2P-based virtual networking, CloudBay achieves scalable resource sharing by disseminating routing information in a decentralized fashion.
- **Self-configurable resource provisioning:** We develop a number of programs to automate network configuration and application deployment in CloudBay. Our work greatly simplifies the task of resource providers and provides timely services to the end users.
- **Fair resource allocation:** A fair allocation of resources allows the service qualities received by end users to be roughly proportional to the costs they pay. In CloudBay, we implement an efficient eBay like matchmaking mechanism to achieve this goal.
- **Flexible resource usage:** CloudBay accommodates a variety of resource usage models and offers differentiated levels of services to end users. For example, it can support both rigid and flexible parallel application execution.

## 4.2   Related Work

There has long been significant interest in investigating the application of economic approaches for resource management in distributed systems. According to Wolski et al. [33], two types of market strategy are commonly used in a computational economy, namely commodities markets and auctions. Auctions are simple to implement and are efficient to sell off computing cycles to contending users. Therefore, auctions achieved wide applications in early computational ecosystems such as Spawn [29], Popcorn [24], and Tycoon [20]. In Faucets [17], auction is conducted to determine the optimal placement of jobs on computational servers. Another early project was Nimrod/G [6], where grid resources were allocated based on user-negotiated contracts with the resource sellers. Most systems were designed for early distributed computing infrastructure such as dedicated clusters and computational grids, and did not account for the latest technology advance in networking and hardware virtualization.

In utility and cloud computing, due to the service-oriented paradigm shift, market-driven distributed systems become commercialized in the next-generation data centers. However, the role of CloudBay is not to serve as yet another IaaS,

PaaS, or SaaS provider, but rather to bridge the scattered HPC resources and the scientific community in support of HPC application development and delivery. The most related work to CloudBay was proposed in [27], where the authors built an experimental resource market inside Google™. The major differences between the two works are presented as follows:

- **Deployment scope:** The resource market created by CloudBay can span multiple networking-layer domains, whereas in [27], the resource market was built upon resources connected by a intra-company network. The capability of traversing NAT/firewalls using P2P virtual networking earns CloudBay a wider deployment scope.
- **Scheduling model:** CloudBay supports immediate service scheduling with transparent job preemption, whereas in [27], auctions were conducted in a periodic manner, and cannot be triggered if the auctioneer did not collect enough bids. As a result, the possible chance of resource utilization during the time of bid window is lost.
- **Pricing algorithm:** In addition to the auction procedure, CloudBay adopts an incentive-compatible payment scheme to regulate bidder behaviors.

Besides the auction approach, many researchers attempted to design incentive-compatible resource allocation mechanisms for individually rational market participants. For example, Teo and Mihailescu [28] developed a strategy-proof pricing scheme for multiple resource type allocations. In [7], Carroll and Grosu designed an online scheduling algorithm MPJS for malleable parallel jobs with individual deadlines. These methods are effective for distributed settings where agents are individual rational and are non-cooperative in a dynamic market.

## 4.3 Design

### 4.3.1 Architecture

Figure 4.1 depicts the architecture of CloudBay. The design goal of CloudBay is to provide a suite of tools that facilitate computational resource sharing and enhance application-to-infrastructure mapping. To fulfill this goal, CloudBay is designed as a service-oriented architecture that seamlessly bridges the gap between applications and resources. First, CloudBay provides resource virtualization services on top of the bare hardware, including: (1) a P2P virtual networking tool that supports scalable and cross-domain resource stitching; and (2) an application-aware VM image called Cloud Appliance that packages grid/cloud computing toolsets and self-configurable networking facilities. With the support of the virtualization service, computational resources residing on different domains can be easily connected together to form an ad-hoc cluster over wide-area networks.

Next, accompanied by the virtualization services, CloudBay offers market-oriented resource-request matchmaking services for both quality-sensitive and budget-sensitive users. The core functionalities include: (1) an account manager

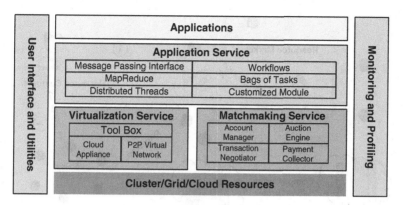

**Fig. 4.1** CloudBay architecture

managing resource seller and buyer accounts; (2) a transaction negotiator that helps to arrange user requests based on the supply and demand level of the current resource market; (3) an auction engine that resolves resource competition when necessary; and (4) a payment collector that collects fees for resource rental. We will cover the details of the market-driven service scheduling scheme in Sect. 4.4.

Finally, CloudBay offers a variety of popular programming models for deploying and running distributed applications. This is achieved by interfacing with pre-packaged software supporting application compilation, run-time configuration and job management. For example, the current implementation of Cloud Appliance image packages MPI library and MyHadoop [19] for HPC application tuning and running. Additional functionalities such as interfacing with users, monitoring and profiling are traversal to the entire CloudBay service stack.

## 4.3.2  Use Case

Figure 4.2 illustrates a simple working scenario in CloudBay. A resource customer submits a bid request (detailed in Sect. 4.4.1) to the CloudBay server seeking to access resources within his budget constraint (step ①). The CloudBay server accepts the bid request and places the request together with other bid requests in the system. If the request cannot be satisfied by the current resource supply, a dynamic ascending auction is launched by the centralized auction engine (step ②). Suppose this user wins 6 VM instances as a result of the auction, the CloudBay server will automatically provide connectivity that bundles the allocated instances into a cluster (step ③). The bundle now becomes invisible to other users and is isolated from other resources in the system. CloudBay employs Condor [11] to manage user submitted jobs, and randomly designates a node within the winning bundle as the head node. The job submitted by the user will be forwarded to a local client node running the *condor_schedd* daemon, and CloudBay will let Condor take over the rest of the work (step ④).

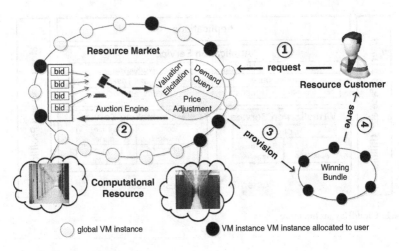

**Fig. 4.2** A simple use case for CloudBay

## 4.3.3   Virtualization

This section introduces our previous work on platform, resource and network virtualization. These techniques form the basis of scalable and self-configurable resource sharing in CloudBay. Since this study mainly focuses on resource management and service scheduling issues, we only present an overview of these virtualization tools, and refer the readers to [15, 30, 32] for implementation details.

### 4.3.3.1   IP-Over-P2P (IPOP) [15, 32]

Just as modern economy is built upon transport infrastructure and freight distribution networks, CloudBay demands scalable and easy-to-deploy networking technologies that support seamless resource stitching and provisioning. CloudBay is designed to provide infrastructure support to scale up to large numbers of geographically distributed resources over wide-area networks. Therefore, in the design of CloudBay, we employ a self-configurable virtual IP network, IP-over-P2P (IPOP) [30], to offer routing capabilities for heterogeneous and self-government peers. IPOP leverages Brunet, a P2P library [30], to unify decentralized computing resources into a ring-structured overlay for routing IP. The development of IPOP is aligned with recent research on virtual private cloud [10, 12, 34]. CloudBay uses IPOP to enable elastic resource provision and relinquish, and attains the following benefits: (1) *scalable network management*, because routing information is self-configured and disseminated over the network in a decentralized fashion. In addition, P2P paradigm efficiently handles node arrivals and departures. (2) *Resilient to failure*, as P2P networks offer more robustness against failure and system dynamics than centralized network management. (3) *Easy accessibility*.

IPOP incorporates a decentralized approach to traverse NAT/firewalls. All these benefits make IPOP a perfect candidate for resource bridging in CloudBay.

### 4.3.3.2 Cloud Appliance

Cloud Appliance directly extends our previous work of Grid Appliance [30]. It packages cloud computing toolsets into an application-aware virtual machine image (available in VMware, Virtual Box and KVM), and supports on-demand resource clustering. A resource provider may choose to launch a Cloud Appliance on the physical host machine, which will automatically place the contributed resource slice into the global CloudBay resource pool. Alternatively, a resource provider may also choose to install separate CloudBay package on the fly (e.g., the package *grid-appliance-base* offers virtual networking functionality and can be installed from Ubuntu). In essence, a Cloud Appliance is an integrated middleware that encapsulates a full job scheduling software stack. *It hides the heterogeneity of various cloud platforms and provides a uniform interface to different cloud resource providers.* Cloud Appliance also allows resource customers to run unmodified, binary software executables without imposing platform-specific APIs that applications must be bound to. Scheduling service in Cloud Appliance directly interfaces with the Condor scheduler for job management. Finally, Cloud Appliance offers sandboxing security such that undesirable behaviors are confined to an isolated VM instance.

Cloud Appliance can be easily deployed on typical x86-based machines. Cloud Appliance can utilize recursive virtualization, which had been supported with hardware support in mainframe hypervisors, but has only recently begun to be supported and its performance implications understood as hardware virtualization in x86 has matured [5]. In our design, the lower-layer virtual machine monitor has the role of encapsulating and distributing the CloudBay stack in a way that makes for simple deployment on a variety of resources, while the upper-layer virtual machine monitor has the role of hosting user's computation. We integrate a set of configuration scripts with Cloud Appliance to ease the process of resource provisioning. When a transaction completes successfully on CloudBay's resource market, Cloud Appliances automatically discover each other and configure themselves into a deliverable resource bundle to serve the job execution request from the winning customer. We have achieved scalable deployment of Cloud Appliances and active user feedbacks from various institutions, as shown in Fig. 4.3.

### 4.3.4 Autonomic Resource Pooling

This section presents the implementation details of resource pooling in CloudBay. Resource pooling involves the development of: (1) a centralized resource pool accessible to all users; and (2) an isolated resource pool allocated to a particular user. Our implementation uses a centralized approach to provide autonomic services for

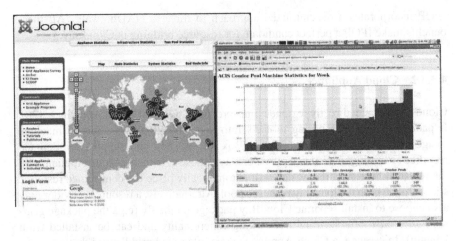

**Fig. 4.3** Snapshot of the deployment of Cloud Appliance. *Left*: A total of 509 IPOP virtual network nodes are distributed across PlanetLab and Archer [2] resources. *Right*: Available number of resources in the virtual appliance Condor pool as nodes are progressively turned on at five different institutions and autonomously join the pool

resource configuration and management. Specifically, a resource manager process, running side-by-side with the Condor central manager, is implemented on the CloudBay server that helps to monitor and manage active resources in the system. We automate the resource joining process by packing a booting script written in Python into the Cloud Appliance VM image. To contribute a VM instantiated by the Cloud Appliance, a resource provider first submits a resource join request from a web interface, and then downloads a certified configuration file bound to the VM. This process is termed as "*floppy insertion*" in CloudBay.

The front end of CloudBay is implemented using Django [13], allowing users to easily interact with the server. When a resource bundle is allocated to some request, the resource manager process will create a new configuration file (floppy) for each VM within the bundle. In our previous implementation [31], users have to manually configure the allocated resource bundle through the web interfaces. Whereas in CloudBay, the resource manager automatically locates the VMs based on their addresses on the IPOP virtual network and transfers the floppies to them via *scp*. This autonomic floppy insertion process enables CloudBay to form an isolated resource pool upon request and greatly simplifies resource allocation.

## 4.4  Market-Driven Service Scheduling

This section presents the design details for user service scheduling in CloudBay. We focus on the eBay-style differentiated service provisioning, HPC job submission, and the design of the auction engine.

## 4.4.1  Model

We consider the resource pool of CloudBay consisting of **dedicated** and **high-performance** computing and storage facilities (e.g., clusters and network shared file systems) that span across organizational and national boundaries. Leveraging techniques presented in Sect. 4.3, these facilities are easy to confederate within a common resource namespace, forming what is referred to as a *science cloud*. Note that CloudBay does not target at non-dedicated and cheap resource in the volunteer computing model because it is hard to guarantee quality of service for quality-sensitive HPC users in a highly dynamic environment. On the resource market formed by CloudBay, resource providers partition their resources into standard sized resource slices that are instantiated using Cloud Appliances, and delegate the task of negotiation and selling to CloudBay. The resource rental model in CloudBay is similar to that used in Amazon® EC2, where users purchase computing services in the unit of instance·hours. However, rather than providing IaaS where users have complete control over the allocated VMs and build their own software stacks, CloudBay is more PaaS-oriented that packs a computing platform and job management functionalities as a service.

Let $\mathscr{R}$ be the set of VM instances within the global resource pool. CloudBay allows for $\mathscr{V}$ classes of VM instances to be created by resource providers (e.g., small, medium and large VM instances). All instances within the same class, i.e., $R_v \in \mathscr{R}, v \in \mathscr{V}$, have homogeneous configurations. We denote the set of user requests by $\mathscr{U}$, each request $U \in \mathscr{U}$ is limited to a set of VM instances within the same class. If a user wishes to run a job on a set of heterogeneous resources, he can simply create a request group in CloudBay that bundles the VMs granted by all the requests sharing the same job configuration.

The need for differentiated service provisioning is imminent because it improves utilization of the infrastructure. Traditional HPC centers allow different job priority classes and use backfilling scheduling [21] to reduce fragmentation of system resources, while modern IaaS providers in cloud computing tend to jointly schedule on-demand and opportunistic resource requests, as is the case of Amazon®'s launch of spot market in addition to the on-demand service. As HPC merges with cloud computing, the question becomes, *how to implement the differentiated request model in modern HPC centers equipped with cloud infrastructure?* In CloudBay, we develop a service scheduling approach inspired by the transaction model used in eBay. Before we proceed to describe our approach, we clarify the assumptions and specifications of the user request model in the next few paragraphs.

CloudBay adopts a market-oriented approach for resource management. In particular, resource pricing in CloudBay is driven by the supply and demand on the market. When user demand is greater than resource supply, resource prices increase that only those resource access requests with sufficient rental prices are satisfied. On the other hand, when user demand falls below the supply level, resource prices decrease that only those resources with sufficiently low leasing prices are allocated. These two cases are symmetric that we can similarly use "sell-it-now" and "bid" options to differentiate different types of resource sellers. In this chapter, we will focus on the case when demand is greater than resource supply.

We define two types of user requests in CloudBay.

- **buy-request**[1]—submitted by quality-sensitive users and is analogous to the option of *buy-it-now* on eBay. The submitted job is likely to be associated with a deadline, and the interruption in service is generally undesirable (non-preemptive). Note that we cannot promise immediate access to the resource because the system might become so congested filling with non-preemptive jobs.
- **bid-request**—submitted by budget-sensitive users and is analogous to those who *bid* on goods on eBay wishing to find a deal. There is no deadline associated with bid-requests. The bidder may specify a expected duration of job execution, or simply let it run to completion. The jobs are characterized as failure resilient that interruption in service does not compromise the computation integrity.

Let $U^b$ stand for a bid request, and let $U^q$ stand for a buy-request, $U^b, U^q \in \mathscr{U}$. A buy-request $U^q$ consists of the number of VM instances to boot, and the expected renting duration. The expression of a bid-request is slightly more complex because it defines flexible configuration parameters. A bid-request is a tuple composed of four elements: $U_i^b = \{v, p_i, n_i, \tau_i\}$, where $v \in \mathscr{V}$ is the requested VM class, $p_i$ is the bid price for a unit VM instance in unit time, $n_i$ is the requested VM number, and $\tau_i$ is the desired renting duration.

Given a mixture of the two types of user requests, the goal of service scheduling is to achieve **fair** pricing while maintaining **high utilization** of the infrastructure. With different context, market fairness could have different meanings. Here by fair pricing we mean: (1) jobs associated with high bids should take precedence over low-bid jobs; and (2) market price of resources is not over- or under-valued. By high infrastructure utilization we mean that the matchmaking service should make resource allocation decisions in a timely manner, and grants resource access rights to end users whenever there is a chance. We will illustrate the design details of service scheduling in CloudBay in the later sections.

### 4.4.2 Job Submission

CloudBay directly interfaces with Condor for job management because of Condor's ability to support both dedicated and opportunistic job execution. We create a uniform web interface that allows users to upload executables and job configuration files to the CloudBay server. The job submission process in CloudBay is illustrated in Fig. 4.4. First, when a resource bundle is allocated to serve a request, the CloudBay server will send the job to a gateway node within the winning bundle

---

[1]The buy-request can be viewed as a special case of bid-request where users are willing to pay a fixed predefined amount.

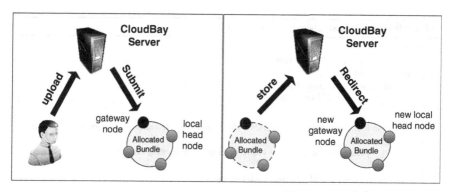

**Fig. 4.4** Job submission in CloudBay. *Left*: A new job submits to an isolated resource bundle. *Right*: A job is preempted from the old allocated bundle and resumed at the new bundle

running the *condor_schedd* daemon. After that, the local Condor server will fetch the job information and schedule the job in the local pool (see the left of Fig. 4.4). If this job gets preempted some time later, the CloudBay server will store the computing state (through checkpointing) as well as the original job configuration. Suppose after a while, a new pool of resources become available again, the CloudBay server will redirect the job information to a gateway node in the new pool to resume the job execution (see the right part of Fig. 4.4).

### 4.4.3   Economy Bootstrapping

CloudBay customers buy computing services using virtual currencies circulated in the system. The recent emergence of the Bitcoin [23] system seems to provide a plausible solution to the implementation of the virtual currencies used in CloudBay. This is because due to the underlying communication infrastructure, the transaction model used in CloudBay is P2P in nature, which matches well with Bitcoin's design principle. For resource providers, CloudBay adopts a closed-loop funding policy [20] to encourage contribution, i.e., each provider is assigned an initial allotment of funds at join time, and earns funds by providing HPC services to resource customers.

### 4.4.4   Service Scheduling

The procedure for request scheduling in CloudBay is summarized in Algorithm 1. In order to eliminate request queueing, the transaction negotiator tries to make an allocation decision whenever a service request arrives. An incoming request issued by some quality-sensitive user takes precedence over all bid requests and gain access

---

**Algorithm 1:** Request scheduling in CloudBay

---

**begin**
   examine incoming request type
   **case** *buy-request*
      **if** *supply ≥ demand* **then**
         allocate VMs as requested

      **else if** ∃ *unfinished bid jobs* AND *their aggregate resource occupation ≥ demand*
      **then**
         preempt jobs from low-bid to high until demand is satisfied
      **else**
         negotiate with the user with two options
          • try at a later time
          • pay large fine for immediate resource access

   **case** *bid-request*
      **if** *supply ≥ demand* **then**
         allocate VMs as requested and collect payments accordingly
      **else**
         start a two-stage Ausubel auction, reconsider bid-requests for all incomplete
         jobs
         allocate according to the auction result

---

to the desired resource bundle whenever possible (lines 3–12). On the other hand, if an incoming request is of bid type, it is scheduled to compete for resources with other bid requests when current resource supply cannot meet its demand. The auction engine will trigger a two-stage Ausubel auction (line 17) to resolve the competition.

The original Ausubel auction (also known as the efficient ascending auction) was proposed in [3], and possesses two appealing properties that make it a good match for our design goal. First, it is *computationally tractable*. Second, it employs a non-linear payment method to *eliminate the incentives of strategic bid behaviors*. However, we cannot directly apply the original Ausubel auction to our scheduling context because of the following difficulties: (1) Ausubel auction uses iterative price adjustment to balance market demand and supply. In practical algorithmic design, the convergence to market equilibrium state might take long time due to price oscillating around the market clearing price. The reason behind this is that it's impossible to determine the step length for price adjustment unless we know the search stop point (the market clearing price) in advance. (2) Some bidders have all-or-none resource acquisition preference. They may suddenly drop out of the auction when price is adjusted. If that is the case, the market equilibrium state may not exist at all. In order to determine resource allocation, we have to extend the feasible region for the solution. Specifically, suppose $n$ bidders bid for $m$ VM instances of certain class $v$. Let each bidder's demand be $d_i^t$ at auction round $t$ (the auction is iterative). We relax the convergence condition of $\sum_{i=1}^{n} d_i^t = m$ to $\sum_{i=1}^{n} d_i^t \leq m$. Note that

---

**Algorithm 2:** The first stage of the modified Ausubel auction

---

**begin**
    sort $p_i^k$ in $\mathbb{L}$ in non-decreasing order
    break ties in $\mathbb{L}$
    `// find market clearing price` $p^*$
    **while** *search range* $> 1$ **do**
        $t \leftarrow t + 1$
        locate medium bid $p^t$ in search range
        **for** *every bidder i* **do**
            $d_i^t = \arg\max_k \{ p_i^k > p^t \}$
        **if** $\sum_{i=1}^n d_i^t > m$ **then**
            shrink search range to the first half of $\mathbb{L}$

        **else if** $\sum_{i=1}^n d_i^t < m$ **then**
            shrink search range to the second half of $\mathbb{L}$
        **else**
            return current bid value as $p^*$
            break

    **if** $\sum_{i=1}^n d_i^t \neq m$ **then** *fail to clear the market*
        `// search for a feasible solution closest to` $p^*$
        backtrack to find the first value making $\sum_{i=1}^n d_i^t > m$
        return the immediately preceding bid in $\mathbb{L}$
    announce winners according to the returned price

---

such relaxation will result in efficiency loss. However, as we use backtracking to find the closest point to equilibrium state, such loss is relatively small.

In the original Ausubel auction, the payment calculation is carried out along with the procedure to search for the market equilibrium price. We propose a *two-stage Ausubel auction* to overcome the first difficulty. In the first stage (summarized in Algorithm 2), the algorithm quickly locates a final market price. With this information, we can decide the price adjustment step and simulate the original Ausubel payment calculation procedure in the second stage. We assume user's valuation to resource bundle is monotonic and strictly concave, i.e., allocated resources exhibit diminishing rewards to users. For a given VM class $v$, let $p_i^k$ (we omit $v$ for brevity of notations) be user $i$'s bid price for the $k$th allocated instance·unit time. This information is collected from the bid submission interface, and is saved in the database of CloudBay's server. To obtain the market clearing price, we perform a binary search on a sorted list of such bid prices. When two bids submitted by two different users tie with each other, the algorithm assigns higher priority to the bid submitted at an earlier wall clock time. If the algorithm fails to converge to a market clearing price, it will backtrack to find the best feasible allocation yielding $\sum_{i=1}^n d_i^t \leq m$. The final allocation for each user is determined by evaluating the marginal bid vector using the returned final market price.

### 4.4.5  Payment Accounting

In the second stage, the auction engine simulates the auctioneer-bidder communications as proposed in the original Ausubel auction [3] in an iterative manner. The payment collector interacts with the auction engine in order to calculate payment amounts for all bidders. We briefly summarize the payment accounting method as follows. First, at each round $t$, the auctioneer calculates the aggregate reserved bundle $\rho_i^t$ for bidder $i$ by comparing the market supply against the aggregate demand from $i$'s opponents:

$$\rho_i^t = \max\{0, m - \sum_{j \neq i} d_j^t\} \tag{4.1}$$

Accordingly, the round reserved bundle $\mu$ is defined as the difference of the aggregate reserved bundle at adjacent rounds:

$$\mu_i^1 = \rho_i^1$$
$$\mu_i^t = \rho_i^t - \rho_i^{t-1} (t > 1) \tag{4.2}$$

Note that $\mu_i^t \geq 0$ because the aggregate demand from $i$'s opponents is weakly diminishing. If $\mu_i^t > 0$, then this amount of allocation is referred to as "clinched" by bidder $i$ at current round price $p^t$. Suppose $i$ wins $A_i$ at the final round $T$, the total payment of $i$ is calculated as:

$$P_i(A_i) = \sum_{t=1}^{T} p^t \mu_i^t \tag{4.3}$$

Accordingly, the auction revenue $\mathscr{Q}$ for the final allocation $\mathbb{A}$ is given by:

$$\mathscr{Q}(\mathbb{A}) = \sum_{i=1}^{n} P_i(A_i) \tag{4.4}$$

One virtue of the Ausubel auction is that it replicates the outcome of the static Vickrey auction. This property is desirable because untruthful users experience degraded performance in computing markets [25]. The proposed auction is incentive compatible (proof detailed in [3]), and results in fair market pricing upon convergence.

### 4.4.6  Discussion

Our scheduling decision is made upon request. This might cause constant thrashing of the low-bid requests. In fact, such an effect is a tradeoff to reduced resource utilization in periodic scheduling. To alleviate this problem, we can compensate the preempted low-bid jobs for a small amount. As the compensation accumulates, the job becomes more resilient to preemption. This is an interesting topic because doing so seems to violate our design goal of fairness. We will explore this issue in our future research.

## 4.5  Evaluation

### 4.5.1  Resource Pooling

We develop and deploy an experimental CloudBay platform composed of 32 VM instances, with 20 of them setup on FutureGrid [14], 8 on Amazon® EC2, and 4 on local lab machines at the University of Florida. Each instance is equipped with 1.5 G memory and 1 virtual CPU core running at 2.66 GHz, and is pre-configured with Condor supporting both dedicated and opportunistic scheduling. The CloudBay server process is implemented and run on a separate machine that also works as the head node for the global Condor resource pool.

First, we examine the setup time for creating an isolated bundle of VM instances. In particular, the setup time is the time elapsed from the moment an allocation decision is made until all resources in the bundle are shown using the *condor_status* command. According to Sect. 4.3.4, the setup time comprises: (1) generating floppy file for network configuration; and (2) notifying the VM instance within the winning bundle about the information of the new Condor head node by transferring the floppy file and modify the local Condor configuration. Figure 4.5 shows the measurement of bundle setup time for multiple VM instances cross three different sites and on FutureGrid only. We observe that the setup time displays an increasing trend as the number of VM instances increases for both experiments. Since cloud users typically request resources over hours, the experiment results indicate that automatic resource pooling in CloudBay imposes a *trivial overhead* to the total resource rental period.

Next, we investigate the performance of CloudBay virtual networking by stress testing the Hadoop cluster with and without IPOP virtual network, respectively. Specifically, we deploy a CloudBay Hadoop cluster, with two VM instances hosted on the UF campus network, and the other two VM instances hosted on Amazon®'s EC2 platform. All instances have the same resource configuration with EC2's m1.large instance type. For the purpose of performance comparison, we also setup a EC2 homogeneous Hadoop cluster connected by EC2's internal network. Two MapReduce programs, *wordcount* and *terasort*, are selected as benchmark programs. For each program, we vary the input file size from 0.5 to 2.5 G, and measure the completion time of all the map and reduce tasks. The results are shown in Fig. 4.6a, b. From the figure, we observe that the heterogeneous networking environment in CloudBay virtual cluster achieves broader deployment scope at the cost of degraded execution time. Using the Hadoop monitoring tool, we observe that

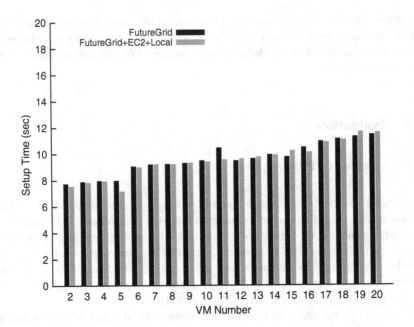

**Fig. 4.5** Experiment for autonomic resource pooling

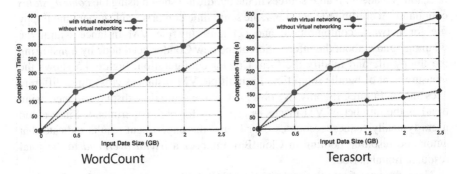

**Fig. 4.6** Performance evaluation for virtual networking. (**a**) WordCount; (**b**) Terasort

the two local nodes greatly straggle the program progress due to the intermediate
data transfer from the EC2 site (the master node is located at EC2 side). In addition,
the performance gap in terms of completion time difference is relatively consistent
in the wordcount program, but increases significantly as the input data size grows.
This phenomenon is primarily contributed to the difference of intermediate data
transfer between the map and the reduce phase. For wordcount, the size of the word
list generated from the map tasks is almost the same for all input,[2] while for terasort,

---

[2]We simply append the same text to generate larger size of input.

the size of the intermediate data for the reduce tasks is increasing all the time. As the data sharing problem becomes more serious in a virtual cloud environment [35], the research for location-aware scheduling mechanism for data-intensive applications is therefore imperative in the future development of CloudBay.

## 4.5.2 Service Scheduling

This section studies CloudBay's ability to schedule mixed-type service requests. Our investigation answers two questions from different perspectives. First, from the perspective of the resource providers, we are interested in understanding how much resource time is consumed by frequent preemptions of low-bid service requests (*preemption overhead*). Next, from the perspective of the end users, we are concerned about the perceived lag of service completion (*service delay*) against the willingness to pay for the service (*offered price*).

The CloudBay platform is in prototyping stage and does not accumulate enough user base. Therefore, our evaluation is simulation-based. We implement a discrete-event simulator using the *Simpy* [26] simulation package based on Python. In the simulation, we create 512 single-core VM instances to serve incoming user requests. Each request can ask for up to 32 instances for running applications. The requested VM number per request is uniformly distributed in the range of $(0, 32]$. For buy-requests, the resource reservation price in a unit of time is set to 20. According to Amazon®'s spot price history [9], we set the offered prices for bid-requests to fall in the range of $(0, 20)$, and follows a normal distribution with $\mu = 8$ and $\sigma = 4$. The job arrival process is assumed to follow a Poisson distribution. By varying the rate parameter $\lambda$, we can simulate system behaviors under different workloads. We generate synthetic user-requested resource usage times based on a realistic workload scenario described in [22]. The workload traces include a Condor workload from the University of Notre Dame, and an on-demand IaaS cloud workload from the University of Chicago Nimbus science cloud. Based on these traces, the requested times are set spanning a relatively long period of time (e.g., a typical request will ask for resource rental over several hours).

In the first set of simulations, we assume the preemption time is linearly proportional to the number of VM instances to relinquish and reset. The preemption process includes the time to save program state (checkpointing), restart the networking configuration process and reconfigure local Condor service. This process can take several minutes for repooling a large number of VM instances. Figure 4.7 shows the results calculated over a 1-month period simulation run. We vary the percentage of bid requests to generate different flows of incoming requests. The labels of *high*, *medium*, and *low* workload correspond to the average system utilization of 83.3, 66.5, and 53.4 %, respectively. Note that the presented results are *relative measurements*, e.g., the bid overhead is measured as the preemption loss with regard to the total resource time occupied by bid-requests, not to resource time occupied by all requests. Therefore, the overall overhead is approximately the

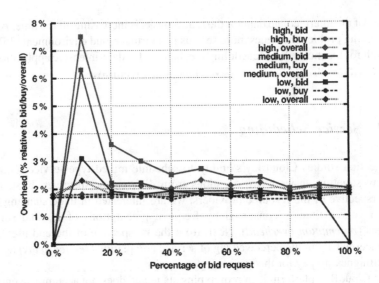

**Fig. 4.7** Evaluation of the overhead due to preemption

weighted sum of bid-request and buy-request overhead. When less bid requests are present, they are subject to frequent preemption by the dominant buy requests. As a result, we observe spikes at the initial phase for bid requests. However, the overall overhead is relatively stable in all the tested scenarios, contributing around 1.8 % to the total busy resource cycles, indicating that CloudBay is suitable for processing high throughput service requests in an open cloud environment.

In the second set of simulations, we create 2,000 synthetic requests and investigate the average service delay with regard to different user bid prices. The *service delay factor* is defined as the ratio of the actual service completion time to the user requested time. A factor of 1.0 means there is no service delay. We conduct five simulation runs with varying percentages of bid request from 30 to 70 %. For each run, the system utilization averages at around 83 %, and the total simulated time is about 50 days. The results are shown in Fig. 4.8. As we expected, higher bid price leads to less service delay in general. However, we also observe a few irregular points on the figure, and the less bid requests a curve gets, the more wrinkled a curve exhibits. This can be explained as follows: (1) a low-bid request might get scheduled without blocking simply because there are available slots in the system; (2) the bid price is randomly generated for each request such that the number of bid requests for a particular price is insufficient. In general, we conclude that CloudBay achieves fair resource allocation for serving differentiated user requests.

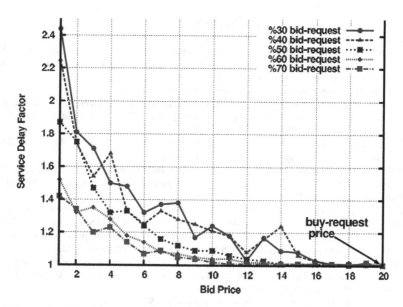

**Fig. 4.8** Service delay factor vs. offered price

# References

1. Abbes, H., Cerin, C., Jemni, M.: Bonjourgrid: Orchestration of multi-instances of grid middlewares on institutional desktop grids. In: Proceedings of the 2009 IEEE International Symposium on Parallel&Distributed Processing (IPDPS'09), pp. 1–8 (2009)
2. Archer. www.archer-project.org
3. Ausubel, L.M.: An efficient ascending-bid auction for multiple objects. American Economic Review **94**(5), 1452–1475 (2004)
4. Bailey Lee, C., Schwartzman, Y., Hardy, J., Snavely, A.: Are user runtime estimates inherently inaccurate? In: Proceedings of the 10th international conference on Job Scheduling Strategies for Parallel Processing (JSSPP'04), pp. 253–263 (2005)
5. Ben-Yehuda, M., Day, M.D., Dubitzky, Z., Factor, M., Har'El, N., Gordon, A., Liguori, A., Wasserman, O., Yassour, B.A.: The turtles project: Design and implementation of nested virtualization. In: USENIX OSDI'10 (2010)
6. Buyya, R., Abramson, D., Giddy, J.: Nimrod/g: an architecture for a resource management and scheduling system in a global computational grid. In: Proceedings of the fourth International Conference/Exhibition on High Performance Computing in the Asia-Pacific Region, pp. 283–289 (2000)
7. Carroll, T.E., Grosu, D.: Incentive compatible online scheduling of malleable parallel jobs with individual deadlines. In: Proceedings of the 2010 39th International Conference on Parallel Processing (ICPP'10), pp. 516–524 (2010)
8. Cirne, W., Brasileiro, F., Andrade, N., Costa, L., Andrade, A., Novaes, R., Mowbray, M.: Labs of the world, unite!!! Journal of Grid Computing **4**(3), 225–246 (2006)
9. Cloud Exchange. Http://www.cloudexchange.org/
10. CloudNet. Available: http://www.cloud-net.org/
11. Condor Project Home Page. http://research.cs.wisc.edu/condor/ (2009)
12. Di, S., Wang, C.L., Zhang, W., Cheng, L.: Probabilistic best-fit multi-dimensional range query in self-organizing cloud. In: Proceedings of the 2011 International Conference on Parallel Processing (ICPP'11), pp. 763–772 (2011)

13. Django. https://www.djangoproject.com/ (2012)
14. FutureGrid. Available: http://futuregrid.org/ (2012)
15. Ganguly, A., Agrawal, A., Boykin, O.P., Figueiredo, R.: IP over P2P: Enabling self-configuring virtual IP networks for grid computing. In: Proceedings of the 20th international conference on parallel&distributed processing (IPDPS'06) (2006)
16. HPC on AWS. http://aws.amazon.com/hpc-applications/ (2012)
17. Kale, L., Kumar, S., Potnuru, M., DeSouza, J., Bandhakavi, S.: Faucets: efficient resource allocation on the computational grid. In: Proceedings of the 2004 International Conference on Parallel Processing (ICPP'04), pp. 396–405 (2004)
18. Krieger, O., McGachey, P., Kanevsky, A.: Enabling a marketplace of clouds: VMware's vCloud director. SIGOPS Oper. Syst. Rev. **44**, 103–114 (2010)
19. Krishnan, S., Tatineni, M., Baru, C.: myhadoop - hadoop-on-demand on traditional hpc resources. Tech. Rep. SDSC TR-2011-2, San Diego Supercomputer Center, University of California San Diego (2011)
20. Lai, K., Rasmusson, L., Adar, E., Zhang, L., Huberman, B.A.: Tycoon: An implementation of a distributed, market-based resource allocation system. Multiagent Grid Syst. **1**, 169–182 (2005)
21. Lawson, B.G., Smirni, E.: Multiple-queue backfilling scheduling with priorities and reservations for parallel systems. SIGMETRICS Perform. Eval. Rev. **29**, 40–47 (2002)
22. Marshall, P., Keahey, K., Freeman, T.: Improving utilization of infrastructure clouds. In: Proceedings of the 11th IEEE/ACM International Symposium on Cluster, Cloud and Grid Computing (CCGrid'11), pp. 205–214 (2011)
23. Nakamoto, S.: Bitcoin: A Peer-to-Peer Electronic Cash System. http://bitcoin.org/bitcoin.pdf (2010)
24. Regev, O., Nisan, N.: The POPCORN marketan online market for computational resources. In: Proceedings of the first international conference on Information and computation economies (ICE'98), pp. 148–157 (1998)
25. Shudler, S., Amar, L., Barak, A., Mu'alem, A.: The effects of untruthful bids on user utilities and stability in computing markets. In: proceedings of the 10th IEEE/ACM International Conference on Cluster, Cloud and Grid Computing (CCGrid'10), pp. 205–214 (2010)
26. Simpy. Available: http://simpy.sourceforge.net/ (2012)
27. Stokely, M., Winget, J., Keyes, E., Grimes, C., Yolken, B.: Using a market economy to provision compute resources across planet-wide clusters. In: Proceedings of the 2009 IEEE International Symposium on Parallel&Distributed Processing (IPDPS 2009), pp. 1–8 (2009)
28. Teo, Y.M., Mihailescu, M.: A strategy-proof pricing scheme for multiple resource type allocations. In: Proceedings of the 2009 International Conference on Parallel Processing (ICPP'09), pp. 172–179 (2009)
29. Waldspurger, C., Hogg, T., Huberman, B., Kephart, J., Stornetta, W.: Spawn: a distributed computational economy. IEEE Transactions on Software Engineering **18**(2), 103–117 (1992)
30. Wolinsky, D., Figueiredo, R.: Experiences with self-organizing, decentralized grids using the grid appliance. In: Proceedings of the 20th international symposium on High performance distributed computing (HPDC'11) (2011)
31. Wolinsky, D.I., Figueiredo, R.: Grid appliance user interface. URL: http://www.grid-appliance.org (2009)
32. Wolinsky, D.I., Liu, Y., Juste, P.S., Venkatasubramanian, G., Figueiredo, R.: On the design of scalable, self-configuring virtual networks. In: Proceedings of the Conference on High Performance Computing Networking, Storage and Analysis (SC'09), pp. 13:1–13:12 (2009)
33. Wolski, R., Plank, J.S., Brevik, J., Bryan, T.: G-commerce: Market formulations controlling resource allocation on the computational grid. In: Proceedings of the 15th International Parallel & Distributed Processing Symposium (IPDPS 2001), pp. 46–53 (2001)
34. Xu, Z., Di, S., Zhang, W., Cheng, L., Wang, C.L.: Wavnet: Wide-area network virtualization technique for virtual private cloud. In: Proceedings of the 2011 International Conference on Parallel Processing (ICPP'11), pp. 285–294 (2011)
35. Zaharia, M., Konwinski, A., Joseph, A.D., Katz, R., Stoica, I.: Improving mapreduce performance in heterogeneous environments. In: OSDI'08, pp. 29–42 (2008)

36. Zhao, H.: Exploring Cost-Effective Resource Management Strategies in the Age of Utility Computing. Ph.D. thesis, University of Florida, Gainesville, FL, USA (2013)
37. Zhao, H., Yu, Z., Tiwari, S., Mao, X., Lee, K., Wolinsky, D., Li, X., Figueiredo, R.: Cloudbay: Enabling an online resource market place for open clouds. In: IEEE Fifth International Conference on Utility and Cloud Computing (UCC'12), pp. 135–142 (2012)

# Chapter 5
# Conclusion and Future Work

**Abstract** We have explored the design space for cost-effective and flexible resource management strategies in utility and cloud computing. In this final chapter, we summarize our findings and discuss related future work directions based on the solutions depicted in this book.

## 5.1 Concluding Remarks

In this book, we have explored the design space for cost-effective and flexible resource management strategies in utility and cloud computing, and proposed a few novel solutions to address the challenges of scalability and heterogeneity. In the second chapter, we investigated the problem of fine-grained resource rental management in utility and cloud computing, and developed solutions for both deterministic and stochastic resource pricing settings. Our optimization models were based on a thorough rental cost analysis of elastic application deployment in the cloud resource market. When resource pricing is fixed, we observed the cost tradeoff between computing and storage emerges in time-slotted resource provision scheduling. Based on this observation, we formulated a deterministic optimization model that effectively minimizes rental cost of virtual servers while covering customer demand over certain planning horizon. In addition, we took one step further to analyze the predictability of spot resource prices using Amazon®'s spot instance price trace, and proposed an alternative stochastic optimization model that seeks to minimize the expected resource rental cost given the presence of spot price uncertainty. Simulations based on realistic settings clearly demonstrated the advantage of the stochastic optimization approach over the predictive approach in rental cost reduction. We also studied the impact of various parameter settings on the performance of both models. We believe the proposed solutions for rental planning offer effective means for resource rental management in practice.

In the third chapter, we presented the management problem of resource trading in a community-based cloud computing environment. The goal of this study is to

H. Zhao and X. Li, *Resource Management in Utility and Cloud Computing*,
SpringerBriefs in Computer Science, DOI 10.1007/978-1-4614-8970-2_5,

investigate the interactions among independent and rational resource traders, and to establish effective and easy-to-implement negotiation protocols for system-wide allocation efficiency and fairness. Towards this goal, we first adopted a multiagent-based optimization framework and analyzed the optimal results without concerning about budget limitation. Next, we proposed a novel directed hypergraph model that combines allocation and envy relationship in a three-dimensional hyperspace. This model effectively captured the impact of trading selection decisions from a global point of view. When budget limitation is imposed, we developed a set of distributed resource trading protocols based on heuristic approaches. Simulation results show that the proposed protocols perform well in a wide range of settings. We expect that the solution for resource trading management presented in this chapter would open new vistas for designing effective resource management strategies.

Finally, we presented CloudBay, a novel resource sharing middleware stack composed of resource management software stack from ground up. Equipped with virtual networking and application-aware virtual appliances, CloudBay achieves ad-hoc self-organization, discovery and grouping of distributed resources without incurring extra deployment and management efforts from both resource providers and end users. Moreover, CloudBay implements a market-driven service scheduling policy that accommodates a mixture of user request models, and efficiently distributes idle resources to users in a cost-effective manner. The pricing and payment accounting policies boosts utilities for multiple parties, and features fair resource allocation for customers. Utilizing services provided by CloudBay, researchers with domain knowledge can comfortably deploy their parallel applications using popular parallel programming models on a resource bundle assembled from multiple organizations. We have already deployed virtual appliances across a variety of open and private cloud platforms, including university clusters, FutureGrid, and Amazon® EC2. We expect that our experiences gained from the design and implementation of CloudBay would open a new research avenue for realizing HPC-as-a-service, and push the boundary for new cloud computing usage models.

## 5.2  A Look into the Future

The research of distributed systems encompasses many areas of computer science and is among the fastest developing fields in the past decade. As resource management needs to cope with the growing complexity of the distributed systems, the exploration presented in this book is just a starting point. We expect the design space to be growing tremendously as distributed systems scale. In particular, this book focuses on the improvement of resource management in distributed systems involving mutually distrustful components. This problem will become more and more important as present and future big data applications call for scalable and reliable computing platforms. The following quote from IEEE Distributed Systems Online [1] published a decade ago has foreseen this challenge, "... *In the past, our approach has been to build systems involving mutually trusting and mutually cooperating subsystems ... We need architectures that support cooperation for achieving a common goal but that do not require subsystems to make strong assumptions about peers"*. Opportunities are emerging to use user-centric approaches that cater to highly dynamic participants. These approaches, such as game theory and auction theory, will inevitably present a substantial body of research for resource management in the years to follow.

Efficiently managing resource allocation is of paramount importance in almost all disciplines of distributed computing. In particular, we are interested in three directions that have the most momentum. The first direction is *Scientific Computing* which applies computational resources to scientific problems. The theoretical peak performance of a single modern GPU has reached 3.7TFLOPS, almost two times as fast as that of the world's fastest supercomputer in year 2000. Such technology advance in computing enables scientists to tackle computing demanding problems with large and costly simulations. Designing efficient resource management strategies for scientific computing is difficult for three reasons. First, uncertainty is ubiquitous in scientific modeling, making resource allocation requirements changing all the time. Second, the need for online processing of the scientific data sets introduces additional demands to computational, storage, and network resource management. Finally, many scientific computing applications involve legacy codes and systems, requiring tremendous efforts to transit to new computing infrastructure. Utility computing and infrastructure clouds offer great potential for scientific users, and we believe the marriage of scientific computing and the cloud will create an exciting perspective in the long run, especially for loosely coupled large-scale HPC applications. This book has presented some of our preliminary research findings on this topic. In the future, we expect to see more research addressing cost and privacy issues of the cloud. In addition, the scientific community needs to invest substantial amount of time and money in developing utility- and cloud-aware tools and services for existing scientific applications and workflows.

The second direction is *Big Data* which deals with high-volume information storage, query and analysis. Interest in big data has given rise to building distributed systems geared for data-intensive processing, and resource management plays a central role in supporting big data applications. In order to handle massive data and meet the performance critical real-time demand, resource management should be agile to allow flexible deployment and provisioning. New platforms have been introduced for big data applications, e.g., Hadoop for better parallelism in computing, and NoSQL for scalable unstructured data storage. The resource management solutions thus need to improve in light of changes in the big data landscape. When evaluating a new resource management solution, performance along with other factors such as deployment complexity, cost, and interoperability with existing solutions, combine to influence the quality of the solution. In this book, we follow this general idea and conduct cost-benefit analysis to resource management in a utility-oriented setting. Another interesting problem is to tradeoff reliability vs. availability of resource allocation, as both resource providers and customers need to determine the allocation of reserved and on-demand resources to minimize waste. Big data applications in cloud also bring more challenge for gracefully handling of resource loss and reallocation. In general, much research and development remain to be carried out to catch up with the ever-increasing pace of data grow.

Finally, as *Mobile and Embedded Computing* proliferate in recent years, efficient resource management is in urgent need to offload heavy computational tasks for mobile and embedded devices. Because these devices are architecturally heterogeneous and resource constrained, they become more and more relied on the cloud computing infrastructure. In order to guarantee Quality-of-Service (QoS) for applications, it is critical to effectively manage resource sharing in data center, and offer more capable network interconnection with enhanced switching and routing. We investigated the economics of resource sharing in this book, and our focus is mainly on the management of computational resource. In the future, we plan to explore management strategies towards network resource sharing. The emerging Software Defined Networking (SDN) technology separates control plan from the data plan and provides centralized control functions with a SDN controller. With this change, it is interesting to examine how multiple network flows belonging to different applications should be shared, scheduled, and priced in modern data centers.

# Reference

1. Milojicic, D.S., Schneider, F.B.: Interview - Fred B. Schneider on Distributed Computing. IEEE Distributed Systems Online 1(1) (2000)